TRENTON HIGH FALLS, NEAR UTICA, N. Y.

SKETCHES OF CREATION:

A POPULAR VIEW

OF

SOME OF THE GRAND CONCLUSIONS OF THE SCIENCES

IN REFERENCE TO

THE HISTORY OF MATTER AND OF LIFE.

TOGETHER WITH

A STATEMENT OF THE INTIMATIONS OF SCIENCE RESPECTING THE PRIMORDIAL CONDITION AND THE ULTIMATE DESTINY OF THE EARTH AND THE SOLAR SYSTEM.

By ALEXANDER WINCHELL, LL.D.,

PROFESSOR OF GEOLOGY, ZOOLOGY, AND BOTANY IN THE UNIVERSITY OF MICHIGAN AND DIRECTOR OF THE STATE GEOLOGICAL SURVEY.

With Illustrations.

NEW YORK:

HARPER & BROTHERS, PUBLISHERS,

FRANKLIN SQUARE.

1870.

TO

THE MEMORY OF THREE SWEET NAMES,

JULIE, STELLA, AND ALLY,

ONCE BLENDED WITH THE LABORS OF THE STUDY AND THE LABOR-
ATORY—THE INSPIRATION OF HIGH AMBITION
AND SUSTAINING HOPE,

NOW ADOPTED IN THE LANGUAGE OF THE ANGELS,

THESE CHAPTERS ARE

AFFECTIONATELY DEDICATED.

PREFACE.

THE work here offered to the public will be found suited, it is hoped, to two classes of readers. There is a numerous class of intelligent persons who do not find it convenient to possess themselves of all the more important conclusions of the physical sciences by a resort either to original memoirs or to formal scientific treatises, but who nevertheless recognize the great interest of the developments of recent science, and would be glad to be put in a position to take a panoramic survey of its grand generalizations. Such an opportunity the author has aimed to present.

The work will also be found useful as an aid in review. The student may plod ever so diligently and ever so intelligently through the details of a science; he is apt to gain only vague impressions and floating ideas, unless enabled to take a comprehensive survey of the field, with the details all left in the background, and the great outlines and prominent landmarks all brought saliently into proper relations to each other. As the engineer, who may have completed the most elaborate survey of a region, requires at last to contemplate it from some elevated hill-top to gain a vivid conception of the landscape as a whole, so the student needs to be lifted up to a position where he may enjoy a bird's-eye view of the entire field at a glance, in order to give vividness, sharpness, locality, and permanence to the thoughts and images floating in his mind.

There also other considerations which have been prominently before the mind of the author in drawing up some of the following chapters for publication. He can not resist the conviction that Nature is intended as a revelation of God to all intelligences. If it be so intended, Nature must be capable of fulfilling the offices of a revelation, and a knowledge of her phenomena and laws must afford the data of a theology. Despite the skepticism of a certain school of recent writers, the phenomena of the universe continue to inspire in the soul of man emotions of religious reverence and worship. To the mass of mind, as to the intelligence of Socrates, and Plato, and Kepler, and Newton, and Galen, and Paley, and Buckland, the order of the Cosmos proclaims an Infinite Intelligence. The author has no fear that the ultimate analysis of the grounds of this belief will result in showing them unreal or unsatisfactory to a critical philosophy. Imbued with such convictions, the author has made no effort to disguise them. He has not, however, entered into any formal attempt to set forth the relations of science to the system of Christian faith, though the way has been frequently opened. He hopes at no distant day to resume the consideration of these subjects. Besides the arguments made familiar by Paley, Whewell, and other writers on Natural Theology — to which, indeed, fourfold strength is added by the later developments of the physical sciences — there are new topics thrust before the world by the current of modern thought, upon which a flood of light is thrown by late discoveries, if, in fact, their discussion does not lie exclusively within the domain of Natural Science. Such are the Antiquity of the Human Race, the Unity of the Race, the Primeval Condition of Man, Harmony of the Mosaic and Geologic Cosmogonies, the Mosaic Deluge, Natural Evil, Development, the Foreshadowing of Man's Birthplace, the Unity of Creation, Teleological and Homological Design in Nature. In the mean time, the suggestions thrown out in this work may be of service to some of those who may be seeking for the grounds of a rational religious belief.

The elucidation of the great problems of philosophic or speculative theology is, indeed, the highest function of science. All our learning would, in reality, be but the "vanity" which it is sometimes reproached with being if it could reflect no light upon the origin, the nature, the duty, and the destiny of man. It is not for its facts, but for the significance of the facts, that science is valuable. To accumulate the data of science is good; to interpret them is the noblest prerogative of a thinking being. Science interpreted is theology. Science prosecuted to its conclusions leads to God.

To all, then, who love to hold communion with the thoughts embodied in the "visible forms" of Nature; who delight to contemplate the sublime, persistent, all-comprehending, and beneficent plans of Deity unfolding through geological cycles toward definite and intelligible ends; in short, to all who love to

"Look through Nature up to Nature's God,"

these pages are respectfully submitted.

THE AUTHOR.

ANN ARBOR, MICHIGAN, *October*, 1869.

CONTENTS.

ILLUSTRATIONS.

SKETCHES OF CREATION.

CHAPTER I.

DISCLOSURE OF THE SUBJECT.

WHAT is this which I have opened from the solid rock? It has the appearance of a bivalve shell, like a clam or an oyster. I was passing a delightful summer-day amid the romantic scenery of Trenton Falls, and broke from the rocky wall of the deep-cut gorge these unexpected forms. Who has not stumbled upon similar shapes at the foot of some beetling cliff, or washed from the weathered soil of some cultivated field? Pause a moment, for these are remarkable and unexpected discoveries. Let us interrogate these forms.

They can not be the shells of oysters or clams; for, in the first place, they are only stone in substance, with a peculiarly dead and mineralized appearance. In the next place, they are nearly three hundred miles from salt water, and as many feet above the level of the sea. Perhaps, then, they are the dead and petrified shells of some fresh-water molluscs, like mussels. This can not be, because the resemblance is not sufficiently close. The beak, or most prominent part of these shell-like forms, is exactly in the middle (Fig. 1, a; see page 14), while the beak of the mussel is always nearer to one end (Fig. 3, a; see page 14). And, farther, one piece or valve of these problematic waifs has a different degree of convexity from the other (Fig. 2), while with mussels both valves are equally con-

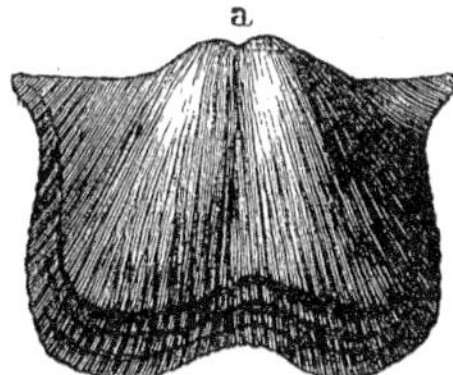

Fig. 1. Fossil bivalve from Trenton Falls; side view of ventral valve. a. The "beak."

Fig. 2. Edge view of the two valves, showing their unequal convexity and depth.

vex (Fig. 4). In fact, the more we study these things, the less they look like mussel-shells—the less they look like any thing else with which we are acquainted. I have heard men familiarly call these objects by the name of "clam-shells;" and others they call "snails;" and still other curious structures, frequently encountered in cultivated fields, they designate as "petrified honey-comb" and "petrified wasps'-nests." But a few moments' careful observation suffices to show that these things differ materially from the objects whose names have been bestowed upon them.

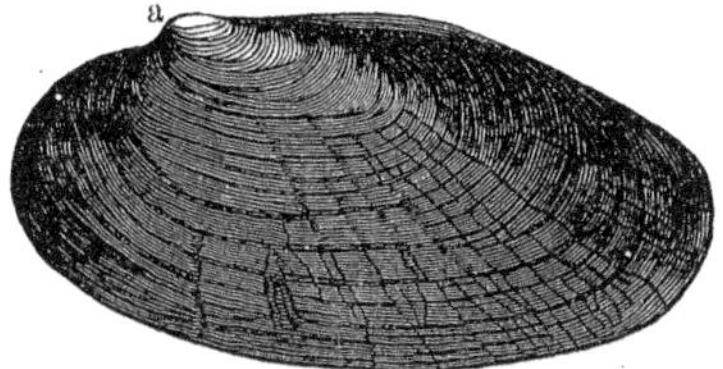

Fig. 3. Common River Mussel. View of left valve. a. The "beak."

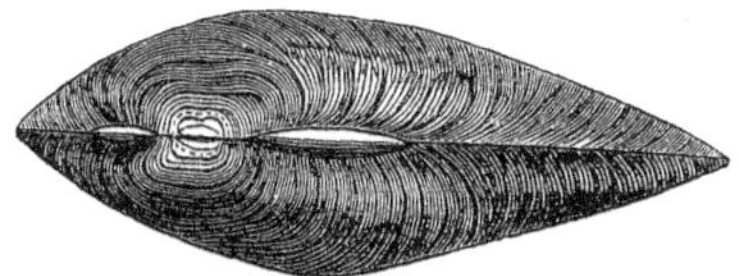

Fig. 4. View of "hinge line" of the same, showing the equal convexity of the two valves.

It seems unreasonable to suppose, therefore, that these shell-like forms have ever belonged to living animals. They are probably but "mere freaks of nature." Perhaps they have been produced by "the influences of the stars."

Or, it may be, there is some mysterious "principle" in the earth which, by some sort of "fermentation," produces these semblances to living forms. Or, still again, as these rocks existed before animals were created, it may be that the Creator moulded these lifeless shapes to serve as "prototypes" or "models" from which the living forms of animals were to be copied. Or, who knows, finally, but the old conjecture of Epicurus may be truth? Since matter must exist in some form, may we not regard these as some of the possible forms under which the particles of matter fortuitously fall?

So reasoned the world prior to the sixteenth century. But this was when the philosopher sat in his closet and argued how things ought to be, instead of going forth to observe how things are. We have learned to contemplate Nature with a different spirit. We have pulled down the house of many a speculatist about his ears. We have demolished many a universe constructed of the cobwebs of logic. We do not despise first principles and necessary deductions, but we have discovered a more direct and a more certain way of arriving at a history of the universe. We interrogate the facts which surround us, and have found them able to narrate a history which never entered the imaginations of the schoolmen. The phenomena of Nature are the premises of our reasoning instead of its conclusions. We have learned to look upon Nature with a profounder respect; and, though the alphabet of our philosophy be trees, and birds, and rocks, and fossils, and other material things which metaphysics affects to despise, we have found that they combine themselves into a language freighted with grand conceptions, and rich in utterances of the unseen, the high, and the holy. It has been revealed to us that the vast system of Nature is the expression of a divine thought—that the wide, blue, restless ocean is the symbol of a divine idea—the swelling prairie, the rocky cordillera, the teeming populations of land, and sea, and air, are the utterances of divine conceptions—the stirring leaf, the basking butterfly, the glistening pebble on the strand, are thoughts of the Infinite, crystallized in visible things, thrown down before us to arrest our attention—strewn over our pathway to provoke our curiosity and arouse the powers of the soul.

We have listened to the recital of the pebble, and its simple story has turned our thoughts backward over the flight of ages, and disclosed a marvelous unity running through the long series of revolutions and innovations to which our domestic planet has been subjected. We have read the epic of the trilobite, and have learned of a Deity inaugurating plans in the infancy of our earth which are still in process of consummation. We have lighted the vistas of the fleeting ages. We have studied the records of universal empires, and the monuments which perpetuate the memory of powerful dynasties. We have seen the procession of living forms pass by, and discovered them marshaled by a single leading Intelligence. We have witnessed the progressive development of the physical world—its successive adaptations to its successive populations, and its completion and special preparation for the occupancy of man, and have learned that the whole creation is the product of one eternal, intelligent *master purpose*—the coherent result of ONE MIND.

What higher subject of contemplation than the world-phenomena which express the thoughts of the Creator? What nobler history to study than the annals of races and revolutions in which the Almighty purpose, instead of human will, has been the controlling power? What antiquities more awe-inspiring than the ruins of continents and the tombs of races whose splendid dynasties passed their meridian and decline while yet the family of Adam was in the unborn future, and God, in the awful solitudes of earth, worked out his all-embracing plans? From the elevated stand-point of modern science, the view before us is inspiring. Let us thread a few of the footpaths leading up to this enchanting altitude.

CHAPTER II.

THE ORDEAL BY WATER.

WE were too hasty in pronouncing it impossible that the little shell struck from the gorge-wall of the roaring stream could ever have belonged to a living animal. It is quite true that no being now exists in the waters of the land or the ocean which can be exactly identified with it. There are forms in the sea, however, which possess every characteristic by which we distinguished it from the river mussel. The resemblances are so close that we are compelled to admit that this may really be a marine form. We look again at the pile of rocks from which this specimen was taken. Layer after layer succeeds from the bottom to the top; and here and there are other similar forms imbedded between the sheets of shale. If these are marine forms, these strata are marine sediments. But here is the difficulty. This place is hundreds of feet above the surface of the sea, and if ever the sea stood at this level the greater part of the continent must have been submerged. But have we not a record of such a submergence? Yes, indeed; the sacred writers tell us of a deluge which destroyed the human family. A tradition of the same has been embalmed in Ovid's myth of Deucalion and Pyrrha; and nearly every nation under the sun, from the cultured Greek to the savage Koloschian of Alaska, has its legend under which the memory of the Deluge has been perpetuated. Shall we then content ourselves with the conclusion that this pile of strata was laid down by the waters of the Noachian flood, and that these molluscs were the contemporaries of the beasts which inhabited the ark? A conclusion thus hastily reached would have suited the preceding age; but the spirit of modern research bids us examine farther. We lay down our little shell, and set out upon the search for evidences to confirm the suspicions already awakened, that it was once the home of a sea-dwelling mollusc.

Go with me first to the coast of the Gulf of Naples. There, near the ancient town of Puzzuoli, at the head of an indentation in the Bay of Baïæ, stand three marble pillars forty feet in height. Their pedestals are washed by the waters of the Mediterranean. The marble pavement upon which they stand, and which was, in the second century, the floor of a temple, or, perhaps, of a bath-house, is sunken three feet beneath the waves. Six feet beneath this is another costly pavement of mosaic, which must have formed the original floor of the temple. What does all this indicate? The foundations of a temple would not be laid nine feet beneath the level of the sea. They must have been built upon the solid land. As the land subsided a new foundation was laid, and a new structure was reared above the encroaching waves. But look upward and examine the surface of the marble. For twelve feet above their pedestals these pillars are smooth and uninjured. Above this is a zone of about nine feet, throughout which the marble is perforated with numerous holes. Exploring these holes, we find them to enlarge inward, and at the bottom of each repose the remains of a little boring bivalve shell—*Lithodomus.* This little bivalve is the same species which is now inhabiting the adjacent waters. We know well its habits. It does not live in the open water. It burrows in the sand, or bores its way into the shells of other molluscs, or into solid stone. But it never climbs trees or marble columns to build its nest, like a bird in the air. How, then,

Fig. 5. View of the Temple of Serapis at Puzzuoli in 1836.

does it occur twenty-three feet above the surface of the water? There evidently has been a time when the whole column, to the height of these *Lithodomi*, was submerged. The oscillations of the surface, therefore, as shown by these indications, were, first, a subsidence and submergence of the original foundation, requiring the construction of the second one six feet above the other; the continuation of the subsidence till the original pavement was twenty-seven feet beneath the surface, at which depth it remained a sufficient time for the little stone-borers to penetrate to the heart of the pillar—a work which they required a lifetime to accomplish. Next occurred an elevation, raising the *Lithodomi* out of the water, and thus ending their existence. Nor is this all. Observations made since the beginning of the present century show that the foundations of this temple are again sinking at the rate of one inch per year.

Such an example, thus authenticated, throws a flood of light upon the problems of geology. It establishes the doctrine of the unstable condition of the land. The rock is no longer the emblem of firmness and stability. We have here a monument which perpetuates the remembrance of secular oscillations in the level of a continent. The little *Lithodomus* has graven the inscriptions upon the marble pillar, even at the cost of its own life. Such care has Providence ever exercised to leave in our hands a key by which to unlock the mysteries of past ages.

It is established, then, that the level of the land may vary—that the shores of a continent may be submerged, and that at a subsequent period they may rise again from the waves. But the doctrine does not rest upon an isolated example. The oscillations recorded upon the temple of Jupiter Serapis are only a clear and beautiful illustration of the nature of the proofs which exist upon every shore.

The columns of other temples are in a similar manner found submerged. Roman roads have been discovered many feet under water. Ancient sea-coasts have been observed far inland. On this continent the shore-line of the Atlantic is experiencing a series of slow, undulatory movements. At St. Augustine, in Florida, the stumps of cedar trees stand beneath the hard beach shell-rock, immersed in the water at the lowest tides. Some of the sounds upon the coast of North Carolina, which have been navigable within the memory of living sea-captains, are now impassable bars or emerging sand-flats. Along the coast of New Jersey the sea has encroached, within sixty years, upon the sites of former habitations, and entire forests have been prostrated by the inundation. In the harbor of Nantucket the upright stumps of trees are found eight feet below the lowest tide, with their roots still buried in their native soil. Similar ruins of ancient submarine forests occur on Martha's Vineyard, and on the north side of Cape Cod, and again at Portland. In the region of the St. Croix River, separating Maine from New Brunswick, the coast has been raised, carrying deposits of recent shells and sea-weeds in one instance to the height of twenty-eight feet above the present surface of the sea. The island of Grand Manan, off the mouth of the St. Croix River, is slowly rotating on an axis, so that, while the south side is gradually dipping beneath the waves, the north is lifted into high bluffs. Near the River St. John is an area of twenty square miles containing marine shells and plants recently elevated from the sea. One hundred and fifty miles east of here, the shore is experiencing another subsidence. The north side of Nova Scotia is sinking, while the south is rising, insomuch that breakers now appear off the southern coast in places safely navigable years ago. The ancient city of Louisburg, on the island of Cape Breton, is another testimony to the uneasy condition of the land. This place was once the strong-hold of France in America, and had one of the finest harbors in the world. It was well fortified, and had a population of twenty thousand souls within its walls. It was destroyed during the French and Indian war, and the inhabitants dispersed. But Nature had herself ordained its abandonment. The rock on which the brave General Wolfe landed has nearly disappeared. The sea now flows within the walls of the city, and sites once inhabited have become the ocean's bed. In 1822 the entire coast of Chili was elevated to a height varying from two to seven feet—an extent equal to the area of New England and New York having been lifted up bodily. In 1831, an island, since called Graham's Island, sprang from the bed of the Mediterranean between Sicily and the site of ancient Carthage. The island is now again but a sunken reef. Another island, as recently as 1866, rose from the bottom of the Grecian Archipelago, before the very eyes of the American consul, Mr. Canfield, bearing upon its slimy back fragments of wrecks that had been sunken in the little harbor of Santorin. Similar ocean-births had many times previously been witnessed in the same vicinity. A hundred and sixty-six years before our era the island of Hyera rose. It was lifted successively higher by earthquake-throbs in the years 19, 726, and 1427. In 1707 Nea-Kameni made its appearance, and in 1773 Micra-Kameni. Even the ancient islands of Santorin, Thrasia, and Aphronisi themselves rose from the sea at the termination of an earthquake some

Fig. 6. View of Graham's Island, July 18, 1831.

Fig. 7. New Volcano of Santorin, 1866.

ages before the Christian era. The ancient Greek fable of the floating islands called Symplegades probably originated in the volcanic movements of the earth's crust in the vicinity of the Thracian Bosporus, the ineffaceable traces of which are still to be seen.

The entire chain of the Aleutian Islands, ranging across the North Pacific from Alaska to Kamtschatka, is but a series of vestiges of an ancient ridge of land now worn out, but originally raised by the power of volcanic fires which are even to-day smouldering beneath the bed of the sea. These fires, as late as 1796, burst out a few miles north of the island of Unalaska, and added another member to the group, which has continued to grow in size till recent times.

As might be expected, the records of continental oscillations are not confined to sea-coast lines, but may be detected along our lakes and in the valleys of the rivers.

If such changes occur in a lifetime, what may not a slow subsidence or elevation amount to in the lifetime of our race? A depression in the valley of the Lower Mississippi of only three hundred feet would admit the waters of the Gulf of Mexico up to the mouth of the Ohio. A trifling depression in Northern Illinois would furnish an outlet to the Gulf for Lakes Michigan, Superior, and Huron. A depression of eight hundred feet would submerge nearly the whole of the Southern and Western States.

How easy, then, in view of facts which every body can observe, to admit the geological doctrine of the former submergence of all the continents. The shells broken from the wall of the gorge at Trenton Falls, though unlike any fresh-water forms, are still the kindred of beings now living in the Atlantic; and, with the evidence before us, we can not resist the conviction that the dominion of the sea once extended over the Empire State. As the relics of Roman dominion are found in England and France, and Germany and Palestine, and nobody questions the testimony of these relics, so the antiquities of Old Ocean have been exhumed from the soil of every state. Who can now perpetrate the folly of denying to one empire the universality which every body concedes to the other?

So reasoned Fracastoro when, in 1517, the exhumation of a multitude of curious petrifactions at Verona, in Italy, had aroused the speculations of numerous writers. But his reasonable suggestion was too bold for the philosophy of that age, and Fracastoro was stamped a heretic by that papal orthodoxy which persecuted also

"The starry Galileo with his woes."

Fig. 8. The work of the Elements at Cape Stevens, entrance of Ward's Inlet, Arctic Ocean.

CHAPTER III.

THE ORDEAL BY FIRE.

IT required a century to gain the credence of the world to the suggestion of Fracastoro. This point gained, it took a century and a half to overthrow the popular belief that the inhumation of fossil remains was all effected at the time of the Mosaic deluge. But few observations of the nature of those already cited had, at this period, been made. With our present knowledge of the oscillations which are going on in the comparative level of continents and oceans, he would seem to be beyond the reach of argument who can still deny that our beautiful prairies have, for ages instead of months, been the bed of a sea which rolled its surges from the Adirondacks on the east to the Sierra Nevada on the west. Admitting the deluge of Noah to have been universal, were the agencies in operation during the one hundred and fifty days of its continuance sufficiently energetic to accumulate sediments twenty miles in thickness in that brief period? Such a conclusion is contradicted by all our observations, instead of being sustained by them. These stratified rocks cover nineteen twentieths of the earth's surface; and the material for them has been ground from the rocky shores of ancient islands and continents by the beating of the waves. If they have thus been distributed by the action of water, it has been a slow process. Admitting, then, the Noachian deluge to have been universal, and to have covered the mountains—since they also are made of fossiliferous strata, even to the altitude of eight thousand feet—is it likely that a hundred thousand feet of sediments would have been deposited in one hundred and fifty days, or at the rate of one eighth of a mile a day?

Consider, also, the myriads of organic remains entombed in these sediments. Their number is fifteen or twenty times as great as that of all existing animals. No evidence exists that the waters of the Mosaic flood were so immensely populous, nor that they were endowed with such destructive energy, as to sweep from existence cubic miles of aquatic forms. And, lastly, it will be noted that four fifths, at least, of the fossil species are now extinct; and, if they were exterminated by the deluge, the objector to geological teaching trips his own feet, for Moses says that Noah preserved pairs of "all flesh wherein is the breath of life, and of every thing that is in the earth." The objector asserts that these animals, now admitted to be extinct, were living at the time of the Deluge, and were exterminated by that event. The sacred historian asserts that the animals living at the time of the Deluge were preserved from extinction by the hand of Noah.

Equally improbable and equally illogical is the position of certain petrified philosophers, who maintain that God created every portion of the earth's crust as we find it. We must thus ignore the indications of every one of a myriad of facts. As well deny that human hands built the Roman aqueduct, or made the pottery exhumed from buried cities or Indian mounds. As well avow our disbelief that Vesuvius ejected the lavas which incrust its sides —that the lightning has struck the riven oak—that the pebble upon the sea-shore has been rounded by the action of the waves—or that the vacated shell by its side was, not long since, the home of an animal enjoying its existence in the brine. Such a belief is to contradict all appearances—to reject that which is most probable and almost demonstrable for that which is contrary to all observation. The geological doctrine is not to deny the unlimited power of Deity, for nothing has done more than geology to unfold and demonstrate that power. It is to apply the same reasoning to geological facts as we apply to other phenomena. In the material world, and within the scope of our investigation, we witness no result which is not the effect of the antecedent energy of what we call secondary causes, operating according to established methods. Whatever can be accounted for by reference to such modes of operation, we feel ourselves precluded from attributing to an extraordinary and miraculous agency.

In view of the facts, therefore, we regard it as certain that a large part of the solid crust of our globe has passed through an ablution in the sea. Particle by particle, grain by grain, pebble by pebble, has been worn from the preexisting rocks; and, after being rolled to and fro for ages by the surges of the sea, has found its way to the deep and quiet ocean-bed. There layers innumerable have been piled upon it. Some of the agencies of Nature have solidified these vast accumulations of sediment; an earthquake-throe has resulted in the birth of a continent, over which the mighty mutations of a geological epoch have swept in grandeur which no human eye was yet created to contemplate; then, in the preappointed order of Providence, man came upon the earth; and to-day, after the lapse of an interval of time which, to human apprehension, is infinite, we split open the solid layer, and behold! the very pebble of granite which was loosed from the primitive rock in the dim ages of the earth's history, which reach far back into the eternity of God! And by its side is a form—an animal form—clearly an animal form; but, if we search the world through, we shall not find its like among existing beings. It is a strange and uncouth form. It was one of

Fig. 9. Shore Erosion on the Mendocino Coast, California.

the earliest representatives of organization upon our globe. Here, in deep ocean solitudes, it lived and sported its day, monarch, perhaps, of an empire thrice the extent of Alexder's. And here it perished—its entire lineage perished. Not a solitary individual has survived; and there is not a living being upon all the earth, or in all the wide realm of the ocean, with which we may compare this anomalous vestige, and determine how it stood related to other beings. Not one, I said; but the faithful explorations of the forgotten zoologist have brought to light, perchance, a solitary family which has inherited the *outré* forms of this primeval ancestor. So do we find the mute monarchs of the ancient continents and seas represented in modern courts by a few obscure individuals still wearing the quaint livery of their antiquated ancestors. Thus do we often witness the remotest past united to the present by single links; and thus do we learn the identity of that Intelligence whose finger-marks remain upon the ruins of past geological epochs, and whose wisdom and benevolence have beautified the landscapes which we daily admire.

But water has not been the only purifier of the materials of the solid crust of the earth. I spoke of "pre-existing rocks," from which the pebble had been broken by the violence of time. These have been purified by fire. Every where do we find these massive crystalline rocks resting beneath the entire series of those which have been accumulated in the form of sediments from water, and which have buried in their common sepulchre the hordes of earth's pre-Adamite existences. These foundation-granites are bearing upon their Atlantean shoulders the weight of twenty miles of solid strata. They contain no organic remains. The granites of this class exhibit no evidence of having been produced from sediments. They bear the marks of fire. The devouring element has caused their stubborn sides to yield. They have been in a molten condition. You may take a fragment and fuse it in a furnace, and, on suffering it to cool under circumstances similar to those in which the rock has been placed, it resumes its rock-like aspect. Marks of heat are all about these granites and their trappean associates. Wherever they have come in contact with rocks of sedimentary origin, the latter are scorched and reddened. In many cases they have been actually fused. A sandstone has been converted into quartz; a shale into a micaceous, semi-crystalline bed; a limestone into statuary marble; and all the vestiges of living forms which these strata inclosed have been withered up and dissipated by the touch of fire.

These underlying crystalline masses are not confined to the deep-seated regions of the earth's crust. We find them thrusting their heads up through the ruptured strata which repose upon their flanks. Higher even than the highest summits formed by the stratified rocks, these foundation masses rear their bold granite heads. From these cold, serene altitudes they look down with dignified complacence upon the fury of the tempest which brings consternation to the landscape below, but fails to ascend to those frigid, breathless summits which every living thing has equally failed to scale.

Some of these venerable domes were reared before ever a particle of sediment had been produced, or even the world-embracing sea had descended from the regions of space around the earth. From their high stations they have watched the procession of all subsequent events; and, while race after race has appeared and disappeared, they have stood calm spectators, unchanged by the myriad vicissitudes of eternity. Others were still the level floor of the ocean when the oldest sediments began to accumulate upon them. In some subsequent age a mighty force has raised them with their load of sediments above the level of the sea. The tempests of succeeding ages have partially stripped them of their sedimentary coverings, and they stand revealed to the light of day. In other cases the tension of the upheaved strata has caused them to break along the crest of a new-formed ridge. A chasm, miles in depth, has opened down to the molten rock below. The fiery sea has risen to the lips of the fissure, and even escaped in a consuming and terrific overflow. In other cases such an eruption has occurred beneath the waters of the sea, and an entire oceanic basin has been converted into a seething cauldron, in which fish and oysters, sea-urchins and lobsters, corallines and sea-weeds, have been cooked together in a Titanian dish of soup. Entire races have thus been exterminated; and, when the elements subsided again to a quiet condition, the waters have been repeopled with countless multitudes of beings exactly adapted to the changed circumstances of the earth—not repetitions of the forms just exterminated, but original conceptions, new ideas from the infinite resources of Nature; and yet not *fundamentally* different, but united to the old by such an identity of fundamental plan as to convince us that the Intelligence which presided in the former epoch survived the catastrophes which brought death to all terrestrial existence, and continued to prosecute his unchanged purposes through succeeding epochs.

Fig. 10. Fingal's Cave in Staffa, composed of basaltic rocks of igneous origin.

Thus fire and water, in their ever-varying operations, have been the principal agencies by which Nature has wrought out the great physical results upon which we gaze with a familiarity which causes us to forget that these safe and solid foundations on which we build cities, and to which we gain a title with hard-earned gold, are but the ruins of pre-existing hills, and valleys, and plains, in which are entombed the long-forgotten relics of the brute nations which preceded us in the possession of the earth.

CHAPTER IV.

THE SOLAR SYSTEM IN A BLAZE.

HAVING made a *reconnoissance* of the vast field which lies before the geological observer, let us ascertain what degree of interest may be derived from a more attentive survey. The ordeal by fire stands first in the order of time. We go back, then, to the molten period of the earth. We plunge into the depths of the past eternity, and behold the terrestial globe glowing with a fervent heat. What a history to trace from that point of time to this! Continents clothed with verdure, and diversified with mountain, hill, and dale—continents spread out upon a thousand courses of solid masonry—are to be derived from this germinal, incandescent mass. It requires an unusual effort of the imagination to leap from the scenes of a modern landscape to an adequate conception of a naked, tenantless, and molten orb, enveloped in an atmosphere of deadly elements, and totally unlike the present earth save in its spherical form and its yearly journey round the sun. To the eye of imagination, the forests must vanish in smoke; the "cloud-capped towers and gorgeous palaces" of man must crumble to clay, and sand, and loam; man and all living things must desert the earth, and leave it in the motionless and stagnant silence of death; the rivers must dry up in their channels; the ocean must change to vapor, and flee to the upper limits of the air; the rock-ribbed mountains must yield to the melting touch of fire; and the rigid crust of the earth must dissolve into a yielding and obedient fluid.

Can we place ourselves in view of the scenes which then existed? Creation is in its incipient stages. The long line of events, which is to end in the installation of man in possession of the earth, lies before us. Methods and plans are now to be adopted whose carrying out is to be extended into the distant future, and which shall comprehend and provide for the endless variety of exigencies which are to grow out of the gradual development of the destined order of things. How inadequate would be a human intelligence to an occasion like this! But to the mind of the Infinite Intelligence the whole creation already existed, and not a feature of the original plan has been abandoned in the long process of its actualization.

But whence the state of things which we are proposing to picture? Was this the "beginning?" In truth, we are forced to admit that science authorizes us to predicate a molten condition of the globe as the consequent of a vaporous one. What are the states of matter but the product of temperature and pressure? We style the liquid the natural state of water, because that is its ordinary condition under our own eyes. But where the mean temperature is below the freezing-point, the solid is its ordinary state; and where the mean temperature rises above the boiling-point, the gaseous is its ordinary state. To men who exist (if such there are) where the climatic temperature never rises to the thawing-point, water is known only as ice; it is quarried as a rock; it may be built into temples, or fortifications, or used for sidewalks. Could man exist, on the contrary, where the climatic temperature never falls below the boiling-point of water, this substance would only be known as a gas, like hydrogen or carbonic acid. There are regions where water, and even mercury, maintain the permanent condition of solids. There are regions where they can exist only as vapor. The pressure remaining constant, the form of these bodies depends upon the temperature. Every one knows that the same is true of sulphur, and zinc, and several other substances. Science has succeeded in changing the form of numerous bodies usually regarded as extremely refractory. Copper, gold, platinum, and the other metals may be readily fused. The same is true of many rocks and minerals. On the other hand, several gases have been liquefied, and some, like carbonic acid, have even been reduced to the solid state. It would seem that, if the appliances of science were as effective as those which we know that Nature wields, every recognized substance might be changed at pleasure into a solid, a liquid, or a gas.

What, indeed, are we to learn from the ejection of melted rocks, in the form of lava, from the throats of volcanoes? Must we not conclude that somewhere within is a reservoir in which all things are melted together?

And what is to forbid our assuming that the history of matter has proceeded, from the remotest epoch to which we can climb, by the chain of cause and effect? What hinders us from mounting *beyond* the molten to the gaseous state of the world? We will do it. We venture to gaze upon a world glowing as an immensity of flame. Matter it must be, but matter in its most attenuated condition. Its preeminent characteristic is *luminosity*. It is primeval light.

But the history of this terrestrial vapor involves the history of the other planets. Geology has become cosmogony. We behold the matter of the solar system—sun, planets, and satellites—but one vast ocean of ignited materials, swung by Omnipotence in mid-space, with other oceans of flaming matter gleaming on it, from every direction, across the cold intervals of infinite space.*

* A period anterior to any definite arrangement of the materials of the earth seems to be mentioned in Gen. i., 1, 2: "In the beginning God

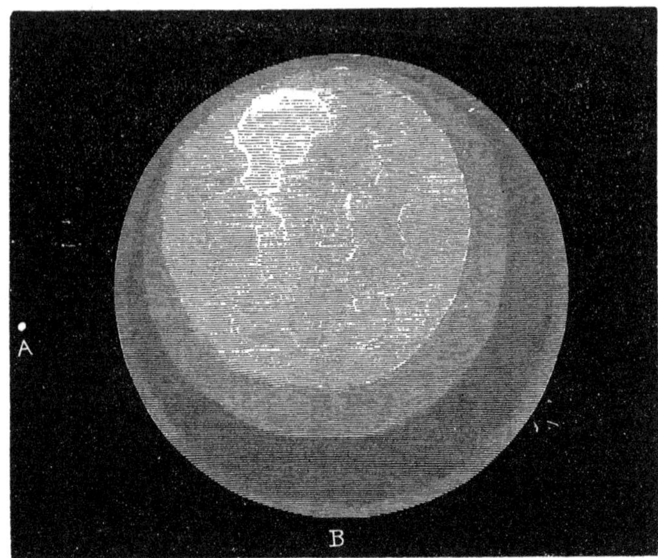

Fig. 11. Comparative volume of the earth in the gaseous and solid state. A. The earth in its present condition. B. The volume of the earth when an igneous vapor.

We dare go no farther; we can go no farther. If science leads us here, she deserts us at this point, and leaves

created the heaven and the earth; and the earth was without form, and void." These interesting utterances will be further considered in another work. A primordial condition of things seems to have been a favorite conception of the ancient philosophers and poets. What a consistent picture is given by Ovid in the "Metamorphoses:"

"Ante, mare et tellus, et quod tegit omnia cœlum,
Unus erat toto Naturæ vultus in orbe,
Quem dixere chaos; rudis indigestaque moles;
Nec quidquam, nisi pondus iners; congestaque eodem
Non bene junctarum discordia semina rerum.
Nullus adhuc mundo præbebat lumina Titan;
Nec nova crescendo reparabat cornua Phœbe;
Nec circumfuso pendebat in aëre tellus
Ponderibus librata suis; nec brachia longo
Margine terrarum porrexerat Amphitrite.
Quaque fuit tellus, illic et pontus et aër;
Sic erat instabilis tellus, innabilis unda;
Lucis egens aër; nulli sua forma manebat."

us to lean only on the arm of Omnipotence. Beyond is only God. No man can predicate an anterior condition of cosmical matter. This condition is necessarily primordial. As matter could not have remained in such a condition—as it did not remain in such a condition—the career of matter must have had a commencement. Its evolutions are not from eternity. As its earliest existence involves an evanescent condition, the *existence* of matter had a commencement. It began to exist only when it began to change. Matter, viewed in the light of physical laws alone, can not be pronounced eternal. Matter is the effect of an efficient cause whose existence is antecedent to matter. As philosophy utters this verdict, how harmoniously rise the voices of the soul, declaring in the face of Atheism that nothing exists except as an effect—demanding that matter itself be remanded to the causation of a creator. And as matter proclaims a First Cause, having existence in itself, as the first link of the long chain of events, so the soul of man reveals an intuition of that First Cause, and rests satisfied in attributing self-existence to a Supreme Intelligence, while impelled to deny it to every thing else.

The beginning of this history does not stretch, therefore, into the inscrutable eternities. We discover the firm Rock of support from which the chain of existence hangs. It is the "Rock of Ages." We feel comforted and strengthened in knowing that "in the beginning *God created.*"

We assert, then, that evidence exists that the solar system came from the hand of the Creator in the state of igneous vapor. Nor does the assertion predicate a condition of cosmical matter that is not, even to this day, exemplified in the universe. Is not the sun a globe of fire-cloud, with a nucleus of molten minerals? And does not the spectroscope declare the composition of the sun to be identical with that of the earth? And what is the substance of the filmy comet that sweeps with such indecent haste through the ranks of the dignified sisterhood of planets? In its dazzling proximity to the sun at perihelion, it can only exist as a fiery vapor, like the substance it seems to be. And if we gaze across the cold and starless interval which separates our firmament of stars from its nearest neighbors, there we may witness a universe in its formative stage. There, indeed, are firmaments so remote that the eye of the telescope is strained in the attempt to descry the component stars; but nearer to our domestic earth than these are the materials of firmaments which remain "rudis indigestaque moles"—the "semina rerum"—the primordial igneous vapor from which worlds are destined to be formed in some far distant future age—so distant, probably, that the career of terrestrial things will first have closed, and mankind will have been ushered into another state of being. Here are specimen creations, postponed to our age in the lapse of eternity, to illustrate before our eyes the infancy of the firmament which is garnished by the nightly splendors of Sirius and Orion. As the gar-pike among animal creations has been perpetuated to our day, to recite the tale of his noble ancestry, so the *Pentacrinus* of the Caribbean still lives to declare the history of pre-Adamite creatures, whose mausoleum is a continent, and the ruins of whose handiwork have risen in mountain piles.

In our attempt to depict the history of this immensity of flame, we draw upon the splendid deductions of Laplace, endorsed by the genius of the elder Herschel, and first foreshadowed by the genius of Leibnitz and Kant. There is every reason to believe that the radiation of heat, which is taking place from the earth and all the planets as well as the sun himself in our own day, is a process which began on the morning of the creation of matter. The rapid loss of heat which the cosmical vapor experienced produced a rapid contraction in volume. Every particle upon the periphery and through the interior began to move toward the centre of gravity of the mass. It is barely possible that a process of cooling and contraction should proceed in such a mass until the work should be completed and no rotary motion be generated. Such a result has, however, in the existing universe, an infinity of chances against it. There were always other masses of matter within our firmament, and others far beyond its limits, which exerted an attraction upon the mass from which our solar system was to be engendered. If Sirius, and Capella, and Vega, and all the other fixed stars, or any of them, be suns like our own, with retinues of encircling planets, their history must be analogous to that of our own system, and we are to regard them as hanging on the verge of the firmament when our system was in its earliest infancy. Their attractive influences were felt. The cosmical vapor which might otherwise have been perfectly spherical became distorted in its form. The position of the centre of gravity was changed. The atoms, in their progress toward the centre of gravity, were found upon lines passing to one side of the centre of gravity. Each began to exert a tangential force. The resultant was a tangential force. It was as if a power had been applied at the surface to inaugurate a rotation of the mass. A rotation once inaugurated in a shrinking globe of matter, it is demonstrable that it would continue to be accelerated as long as the mass should continue to contract. In the present case the mass assumed the form of a greatly flattened spheroid, and the velocity of the peripheral portion became so great as to overcome the power of gravity. As a consequence, the peripheral portion became detached in the form of a ring—as water is thrown from a rapidly revolving grindstone. The ring continued

Fig. 12. The Solar System rotating in a gaseous state.

its rotation about the mass till the oscillations to which it was subjected produced a rupture, when the whole material of the ring gathered itself together in another globe of igneous vapor revolving around the first.

In progress of time the principal mass, under the influence of inevitable refrigeration and acceleration of its motion, threw off another ring, which, on rupturing, became another revolving globe. From time to time the process was repeated; and a series of globes was thus left at varying distances from the centre of the system. These globes became the planets, and the residual mass is the sun. We come into existence, and gaze upon the series of planets, on one hand, and the sun upon the other, and think, because no perceptible change transpires in a generation or two, that all things are stable—that creation is completed—that all things were made at first as we see them, and are destined so to remain. Vain thought! The movements of matter are even now in progress. The residual mass—the sun—is still cooling and shrinking, and may yet throw off other rings, the germs of other planets within the orbit of Mercury—if, indeed, Lescarbault be not correct in asserting the existence already of an intramercurial planet.

But what of the detached globes of matter? The largest are the remoter, being formed of rings detached when the parent mass was largest. Each has continued to revolve in an orbit which marks the periphery of the parent mass at the time of the planet's separation. All continue to revolve in the same direction as the parent mass and the resultant sun. All revolve very nearly in the plane which must always have been the plane of the equator of the mass—the astronomical ecliptic. All continue to revolve upon their own axes in the same direction as required by the motion of the parent mass. Can all these things be so by chance? Can these planetary movements thus correspond, and the material constitution of all these bodies be identical, without leaving a profound conviction upon our minds that they have had a common origin and a common history? Such queries were raised by Leibnitz and Kant upon slenderer data than we possess. Does not the hypothesis of Laplace rise almost to a demonstration?

But what, again, of our family of infant planets? Each sprang forth a globe of igneous vapor like their common mother. Each began to repeat the process of cooling, condensation, and accelerated rotation. In the cases of the larger, the cooling had not reached the point of liquefaction before the rotation had become sufficiently rapid to detach from one to six or seven rings, which, in turn, became satellites revolving about their planets. The larger planets have had time to detach the greater number of rings. Our earth threw off but one, and became too rigid to repeat the process. Mars, Venus, and Mercury—all smaller than the earth—attained the rigid condition before their acquired velocity had separated the periphery. Their nights are consequently unillumined by the presence of a moon. Saturn not only threw off seven rings which became satellites, but another also, which to this day hangs poised in a state of unstable equilibrium—as if the hand of Omnipotence had steadied it, and arrested it in its career, to hold it up to the gaze of intelligent creatures, to reveal to them the nature of events which transpired before their arrival upon the theatre of existence. And this ring is said to be a liquid—a discovery for which we are indebted to the analysis of an eminent American scholar, but one which lends still farther corroboration to our view of the genesis of worlds.*

We have then, preserved as if by the care of Providence,

* The only difficulty arises from the fact that the liquid ring is not self-luminous. But this difficulty is not insurmountable. It may be aqueous.

existing exemplifications of all the main phenomena which have attended upon the evolutions of cosmical matter from the time when it first sprang from the hands of its Creator. The cloud-like comet; the "zodiacal light;" the solar-photosphere; the irresolvable nebulæ, may probably be regarded as examples of attenuated luminous matter such as our theory hypothecates. Every whirlpool shows how rotation is liable to be spontaneously generated. The Saturnian ring or rings illustrate an essential phase in this cosmical genesis; their liquid condition another. The body of the sun is a mass remaining in the incandescent state; while the planets have become opaque, because smaller masses of matter sooner reach the point of total refrigeration. The moon represents a state of refrigeration which the earth is destined to attain in the distant future. We may thus regard the visible universe as a vast museum in which Nature has preserved for our instruction specimens illustrative of every stage in the embryology of worlds.

Will it be asked how such views accord with our theistic opinions? I reply, perfectly. It has become a kind of fashion in certain quarters to denounce all scientific doctrines to which the much-abused term "development" can be applied; but in this we may be too much influenced by "the fashion." Leading theologians—though indeed scarcely followers of the leaders in physical science—have heaped opprobrium on the "nebular hypothesis" as tending to atheism. The patronage of this hypothesis by the author of the "Vestiges of Creation" has thrown a dark suspicion over it; but the cause of truth will be best promoted by allowing every scientific question to be decided on its merits. The scientific world as a whole will never abandon a position because denounced by the theological world—not even because it seems to be in conflict with sound theological doctrine. Scientific evidence is of such a nature as always to command the respect and the assent of the bulk of reasoning men. If this hypothesis is sustained by scientific evidence, it is the duty of the Christian world to embrace it and convert it to their own uses. To do otherwise is to earn the contempt of those who are really on the side of truth. If it is not sustained by scientific evidence, it behooves the Christian world to overthrow it from scientific data. Such data ought not to be monopolized by secular learning. Science belongs peculiarly to Christianity, and Christianity is in duty bound to assert her claim. If she can use science to overthrow a false and dangerous position, she is derelict to neglect the opportunity; and all her denunciation will not atone for the error.

But this hypothesis, whether it represent the true history of cosmical matter or not, has no tendency to remove the Deity from creation. This has been admitted by Whewell, Buchanan, and all others who have been crowded to a response. This objection is founded in short-sightedness and a failure to appreciate the case. The hypothesis simply assumes that the Creator has brought worlds into existence by the use of second causes, precisely as he brings a tree into existence. Does any one hesitate to admit that an oak has undergone a slow and regular "development"—or that the delta of the Mississippi is undergoing "development"—or that the cone of Vesuvius is undergoing "development?" If it appear to intellects of the loftiest and broadest grasp that the Creator has evolved the solar system according to a method, and by the use of natural laws, exactly as he evolves a tree from the germ in the seed, why do we charge atheism in the one case and not in the other? The only difference between the cases is that the one attributes to Deity a vaster scope of intelligence and power than the other—and in doing this it concedes to him more of that character which constitutes him God. In short, it is the more theistic view of the two.

This hypothesis has also the merit of dating the commencement of the evolutions of matter—which to some extent all must admit—back to the very point beyond which it is impossible for science to predicate any thing except to drop the universe into the hands of a First Cause. It places science in the position where, instead of suggesting a query or doubt, she naturally, and inevitably, and cheerfully pronounces the name of God.

CHAPTER V.

THE REIGN OF FIRE.

WHATEVER may be thought of the evidence bearing upon the question of the former gaseous condition of our world, or of the entire solar system, it is generally admitted that the evidence of former igneous fluidity is somewhat conclusive. This is a doctrine which we may regard as resting on legitimate geological data. This is a condition of the world we may proceed to contemplate without serious misgivings. Our earth was once a self-luminous star.

At the temperature which would fuse the mass of the rocks, all the more volatile substances could only exist in the form of an elastic vapor surrounding the earth. All the carbon in the world must have existed in the form of carbonic acid; all the sulphur as sulphurous acid; all the chlorine as chlorhydric acid; all the water as an invisible elastic vapor, extending out beyond the limits of the present atmosphere. There could hence be upon the earth no vegetation, no animals, no limestone, no salt, no gypsum, no water. All that we now behold must have been represented by a glowing, liquid nucleus, enveloped in a dense atmosphere of burning acrid vapors. This orb, by the immutable laws of physics, must have revolved upon its axis and performed its revolutions around the sun. The sun and moon (if the latter existed) must have raised the fiery ocean to a tidal wave which rolled around the globe—the type of an action which has continued to the present period. There were also day and night. The sun rose in the morning, and sent a lurid ray through the dense, refractive atmosphere, and at night sank into the smoke that ascended from a burning world. The morning and evening twilight almost met each other in the midnight zenith, so high and so refractive was the heterogeneous atmosphere. But there was no need of twilight. An ocean of fire sent up to the nocturnal heavens a glare that was more fearful than the poisoned ray of the feebly-shining sun. Here was chaos. Here was the death and silence of the primeval ages, when the Uncreated alone looked on, and saw order, and beauty, and life germinating in the heart of universal discord.

In obedience to the law of thermal equilibrium—a law which undoubtedly rose into being with the birth of matter—the high temperature of the earth gradually subsided through radiation into external space. A crystallization of the least fusible elements and simple compounds eventually took place in the superficial portions of the molten mass. This process continued till a crystalline crust had been formed, resting upon the liquid mass which still constituted the chief bulk of the globe.

It has sometimes been objected to this view that the solidified materials would possess superior density, and would, accordingly, sink into the liquid portions. If this were so, the solidification of such a molten mass would either commence at the centre, or a uniform refrigeration would proceed till the whole would suddenly be consolidated. It is the general belief that the central portions of the earth still remain in a molten condition, while the habitable exterior is but a comparatively thin crust. [See Appendix, Note I.] If this belief is well founded, the first solidified portions did not descend toward the centre. Moreover, we know that, in the case of water and several other substances, the newly-solidified parts are less dense, and float upon the liquid portions. This apparent exception to the law of expansion by heat is accounted for by supposing that, when the molecules of a solidifying fluid arrange themselves in a regular crystalline manner, they inclose certain minute spaces, so that the resulting crystal is a little more bulky than the unarranged molecules from which it was constructed. And this may be the case, even though a cooler temperature has caused them to shrink into closer proximity (for they are never in contact) than before crystallization. If this law applies to the refrigeration of water, type-metal, iron, and other substances, we may reasonably infer it to be a general law of matter. We should expect, then, that crystals of quartz would float upon molten quartz, or solid trap upon molten trap, just as solid iron floats upon molten iron, or solid ice upon molten ice. We have, therefore, not only evidences of the fact of a forming crust, but also a probable means of accounting for it.

We may conclude, then, that a solid film began to form over the surface of the molten sea. But the earth was even then, as from the beginning, obedient to the law of axial rotation; and the sun and moon reached forth, with their attractive influences, to solicit the mobile rocks into tidal elevations. As the wave pursued the moon around the earth, it daily ruptured the forming film, and only a wilderness of floating fragments remained, strewn over the surface of the fiery abyss. In due time, however—let us be liberal in our concessions of time—the rocking and jostling fragments became permanently frozen together, as the broken ice of Arctic seas, after being worried by winds and currents, seizes an interval of calm to consolidate into a vast and rugged floe. So the rock-floe of this fiery ocean formed, at length, a bridge of rough and sturdy strength. It was a mixed conglomerate of crystalline fragments, such as we now witness in some of the *granites*, which are mixtures of quartz, feldspar, and mica; or the *syenites*, which are mixtures of quartz, feldspar, and hornblende; or the *diorites*, which are mostly mixtures of feldspar and hornblende. Or, perchance, the solidification took place under such circumstances that the crystallization was more obscure, as in the various *dolerites*, which every one admits to have been born of fire. We say that the process of refrigeration must have resulted in such rocks as these; and it is a curious and instructive fact, that when we turn our attention to an examination of the oldest rocks, we find granites, and syenites, and diorites, and dolerites resting where we expected them, underneath the rocks that came into being after water existed upon the earth, spreading out their bases in every direction, and constituting the very abutment which supports the lithological pile. We thus trace a certain succession of events which must occur in accordance with the established laws of physics, and find the series of sequents confirmed by the facts of the rocks themselves. Though this mode of reasoning is not in the spirit of modern natural science, it must always lead us to the truth if we reason correctly. Nevertheless, it is seldom the case that we are justified in the attempt to predicate the phenomena from the laws which involve them, as long as it is our privilege to confirm the laws by a study of the phenomena. In the present instance, the history of science shows that the laws were first arrived at by a careful induction from facts; and the little deductive reasoning in which we have indulged is but tracing the thread a little farther back, with the phenomenon it hangs upon all the time in full view.

In the process of refrigeration the stiffening crust would become too large for the nucleus within. This would necessarily result from the more rapid contraction of the more highly heated portions. If the solid and the molten portions suffered equal losses of heat, the molten, by shrinking the most, became too small for the enveloping crust. The crust, therefore, must wrinkle, to fit the shrinking nucleus. Thus incipient inequalities of the surface began to appear. These were the germs of mountains and of continents. From a new-born wrinkle grew the lofty Cordillera.

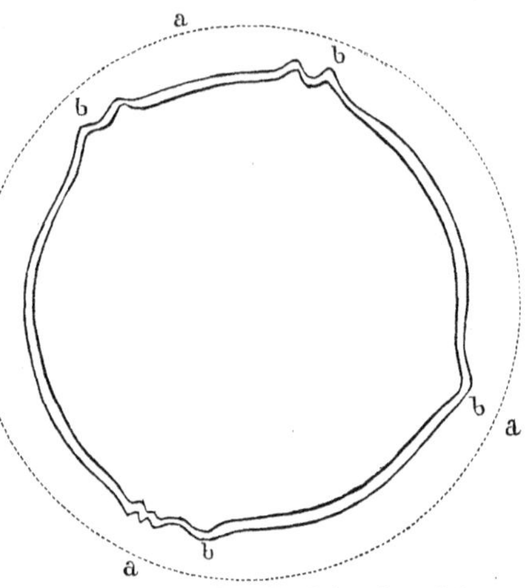

Fig. 13. Ideal Section of the Earth in primeval times. *a, a, a.* The surface when solidification first commenced. *b, b, b.* Wrinkles developed in the crust by the shrinkage of the nucleus.

A scene of terrific sublimity approaches. As yet no water existed upon the earth. No rain had fallen upon the parched and blackened crust. All the water which now fills the oceans, and the rivers, and the lakes—all which saturates the atmosphere, and the soil, and the rocks—rested then upon the earth as an arid, elastic, invisible vapor, extending an unknown distance into surrounding space. This vapor was not cloudlike, but intensely hot and transparent. It was a gas, like the steam just issuing from the escape-pipe of a steam-boiler. The time had now arrived, however, when the remoter regions to which this aqueous gas extended began to be so far reduced in temperature as to cause condensation to begin—as the heated steam, rushing from the locomotive, soon cools into a cloud of visible mist. An intelligence located upon our earth at this epoch would have seen the dusky atmosphere begin to thicken. In the far-off regions, wisps of vapor crept along the sky, as cirrhi in our day foretoken the gathering storm. They grew, and thickened, and darkened till a pall of impending clouds enwrapped the earth, and the light of sun, and moon, and star was shut out for a geological age.

Particle drew particle to itself, and rain-drops began to precipitate themselves through the lower strata of the fervid atmosphere. In their descent they were scorched to evaporation, as the meteor's light vanishes in mid-heaven. The vapors, hurrying back to the bosom of the cloud, were again sent forth, again to be consumed. At length they reached the fervid crust, but only to be exploded into vapor and driven back to the overburdened cloud, which had an ocean to transfer to the earth. The clouds poured the ocean continually forth, and the seething crust continually rejected the offering. The field between the cloud and the earth was one stupendous scene of ebullition.*

But the descent of rains and the ascent of vapors disturbed the electricities of the elements. In the midst of this cosmical contest between fire and water, the voices of heaven's artillery were heard. Lightnings darted through the Cimmerian gloom, and world-convulsing thunders echoed through the universe.

> "The sky is changed! and such a change! Oh, night,
> And storm and darkness!"

* Those who are acquainted with Figuier's interesting works will note a remarkable correspondence between his treatment of this subject and my own. It is but justice, therefore, to state that these chapters were drawn up long before the work of Figuier appeared. This, indeed, has been my conception of these primeval scenes since 1856; and it was in print in 1857.

Fig. 14. The Primeval Storm.

CHAPTER VI.

OLD OCEAN COMMENCES WORK.

A THOUSAND years of storm and lightning have passed, and the primeval tempest is drawing to a close. The waters are now permitted to rest upon the surface. By degrees the clouds are exhausted, and sunlight filters through the thinned envelope. As the morning of another geological epoch dawns, it reveals the change of scene. The surface which, in the preceding age, was scorched and arid, is now a universal sea of tepid waters. The earliest ocean enveloped the earth on every hand. A few isolated granite summits perhaps protruded above the watery waste. Around their bases careered the surges which gnawed at their foundations. Geology is unable to aver that any of them survived the denudations of this first detrital period. The demands of nature for material from which to lay the thick and massive foundations of the stratified pile of rocks were enormous, and it is probable that whole mountains were quarried level by the energies of this young, fresh, and all-embracing ocean. Probably, however, the nuclei of some of our oldest mountain masses, though subsequently elevated to their present altitudes, may be regarded as the remnants of the granite knobs that reared their frowning and angular visages above the primordial deep. If so, the erosion of the waves and the battering of the tempests have given to their sides and heads a smooth and bald rotundity. But most, if not all of the original pinnacles of the earth's crust have been leveled to the water's surface and spread over the floor of the sea. To-day we may gather up the fragments, not from the bottom of the sea, but raised again mountain high, or incorporated into the fabric of new-built continents! Sublime ruins! What are the marbles of Nineveh, or the columns of the Parthenon, in comparison with these hoary relics of Nature's primeval structures?

I said that the fury of the waves strewed the ocean's bed with the ruins of these ancient islands. This is no fancy. The demonstration is before our eyes. The floor of the sea was first formed of rocks that had cooled from a state of fusion. The few islands that existed were but exposed portions of this floor. The *débris* scattered over this foundation would be arranged in layers, as water always arranges its sediments. The coarser materials would be transported by the more powerful action and deposited in one place; the finer materials would be carried beyond by

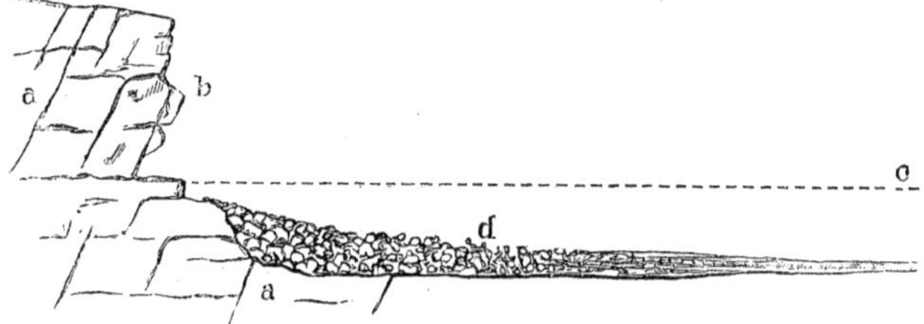

Fig. 15. Shore Erosion and Distribution of Sediments.

a, a. The primordial igneous crust. b. A sea-side cliff gnawed by the waves. c. The ordinary sea-level. d. The ruins of the cliff—the coarser deposited near the shore, and the finer floated to greater depths.

the feebler agency, and deposited in a remoter region. Thus some of the first-formed strata would be finer and others would be coarser; but all must be composed of materials derived from the pre-existing rocks. This deduction is again corroborated by well-known facts. Every where do we find reposing upon the ancient igneous floor a bed of stratified materials composed of the same constituent minerals as the rocks they rest upon. For instance, granite is very commonly the foundation rock; but immediately upon this repose thick beds of gneissoid rocks. Now gneiss, like granite, is composed of quartz, feldspar, and mica, and differs only in this—that the constituents have been broken up, assorted by water, and redeposited in regular layers. As we have different varieties of granitoid rocks, so we have corresponding varieties of gneissoid rocks, differing from the former only in being stratified. So general and so well recognized is this phenomenon, that Sir Roderick I. Murchison, an eminent geological authority, designates these lower strata beds of "fundamental gneiss." This occurrence of gneiss, every where reposing upon granite, is a most interesting and instructive fact, and confirms all that I have said of the denudation of the primitive islands, and the universality of the primitive sea.

But, though gneiss is generally the foundation stratum, we find abundance of other rocks either reposing upon the gneiss, or interstratified with it in the lower portions of the sedimentary series. Undoubtedly some of these have resulted from the impalpable powder to which long-continued attrition reduced some portions of the primitive granite, transported to the remotest and quietest portions of the ocean, and there allowed to subside. But we know also that others of the oldest strata associated with the gneisses have been the results of chemical agencies. This is one of the revelations of modern chemical geology, which no name has more adorned than that of Dr. T. Sterry Hunt, of the Geological Commission of the Dominion of Canada. According to Hunt and Logan, the *limestones* of this early period could have had no other than a chemical origin. Common limestone is composed, as every one knows, of carbonic acid and lime. Heat, as the manufacturer of lime illustrates, expels the carbonic acid in the form of a gas. Under the high temperatures of the earliest periods, therefore, limestone could not exist. It has already been stated that all the carbon, sulphur, and chlorine in existence must, in those periods, have been represented by carbonic (CO^2), sulphuric (SO^3), and chlorhydric (HCl) acids, existing in a volatile state, mingled with the other gaseous constituents of the atmosphere. At the same time, all the silica of the globe, playing the part of an acid, would unite with the fixed elements, producing silicates of complex constitution —just such silicates as we actually find entering into the structure of the oldest portions of the earth's crust. The first rains which descended would be charged with the atmospheric acids just mentioned, which, attacking the solid silicates at a high temperature, would, as the analytical chemist knows, produce reactions resulting in the chlorids of calcium (ClCa), magnesium (ClMg), and sodium (ClNa), mingled with the sulphates of these bases (SO^3KO, SO^3NaO, SO^3CaO, SO^3MgO). The liberated silica (Si^2O^3) would separate, and would be chemically precipitated during the subsequent cooling of the waters, and would thus give rise to the enormous beds of quartz which we actually find among the very oldest strata, but nowhere else.

Among the other silicates originally formed is a family of minerals known as feldspars—very abundant, and containing, besides alumina, large percentages of either potash, soda, lime, or lithia, or two of these alkalies together. The decomposition of these feldspars—especially orthoclase, or potash-feldspar ($Si^2O^3Al^2O^3KO$)—must have taken place on an extensive scale. The result would be a clayey hydrate, called kaolin ($Si^2O^3Al^2O^3$) when pure, which became the basis of many clays and other argillaceous rocks like graphic and roofing slates. The remainder of the orthoclase would be in the form of silicates of potash (Si^2O^3KO) and soda, which would remain in solution in the sea. But the carbonic acid of the atmosphere, having a more powerful affinity for these alkalies than the silica, would wrest them from combination with the silica, as already stated, and would form carbonates of potash (CO^2KO) and soda (CO^2NaO), while the silica would be added to the quartzose rocks of the globe. These carbonates, whether formed in the ocean or on the hill-sides, would, when transported to the ocean, find themselves confronted with chlorid of calcium (ClCa), and probably other chlorids. Chlorid of calcium, carbonate of potash (CO^2KO), and carbonate of soda (CO^2NaO), brought face to face, would immediately enter into arrangements for an exchange of partners. Carbonic acid (CO^2) would incontinently abandon potash (KO) and soda (NaO), and betake itself to calcium (Ca), changing its name, by the aid of a little oxygen, to "lime" (CaO), and forming a union known as carbonate of lime (CO^2CaO). With equal celerity, chlorine (Cl), dispossessed of its calcium (Ca), would compensate itself by seizing upon potash (KO) and soda (NaO), and, after eliminating the oxygen (O) in their constitution, would unite with potassium and sodium, forming chlorid of potassium (ClK) and chlorid of sodium (ClNa). Thus all parties would be better satisfied, and each would abide in its appropriate place. Carbonate of lime (CO^2CaO) refusing, for the greater part, to be dissolved in sea-water, would settle to the bottom and become limestone; while chlorid of sodium (ClNa)—which is only the chemist's name for "common salt"—remained in solution, and thus gave its characteristic salinity to the sea. Chlorid of potassium (ClK) also continues to exist in sea-water in smaller quantity.

The diagram on the following page is intended to represent to the eye the chemical reactions above described. The symbols are familiar to the chemical reader; but they

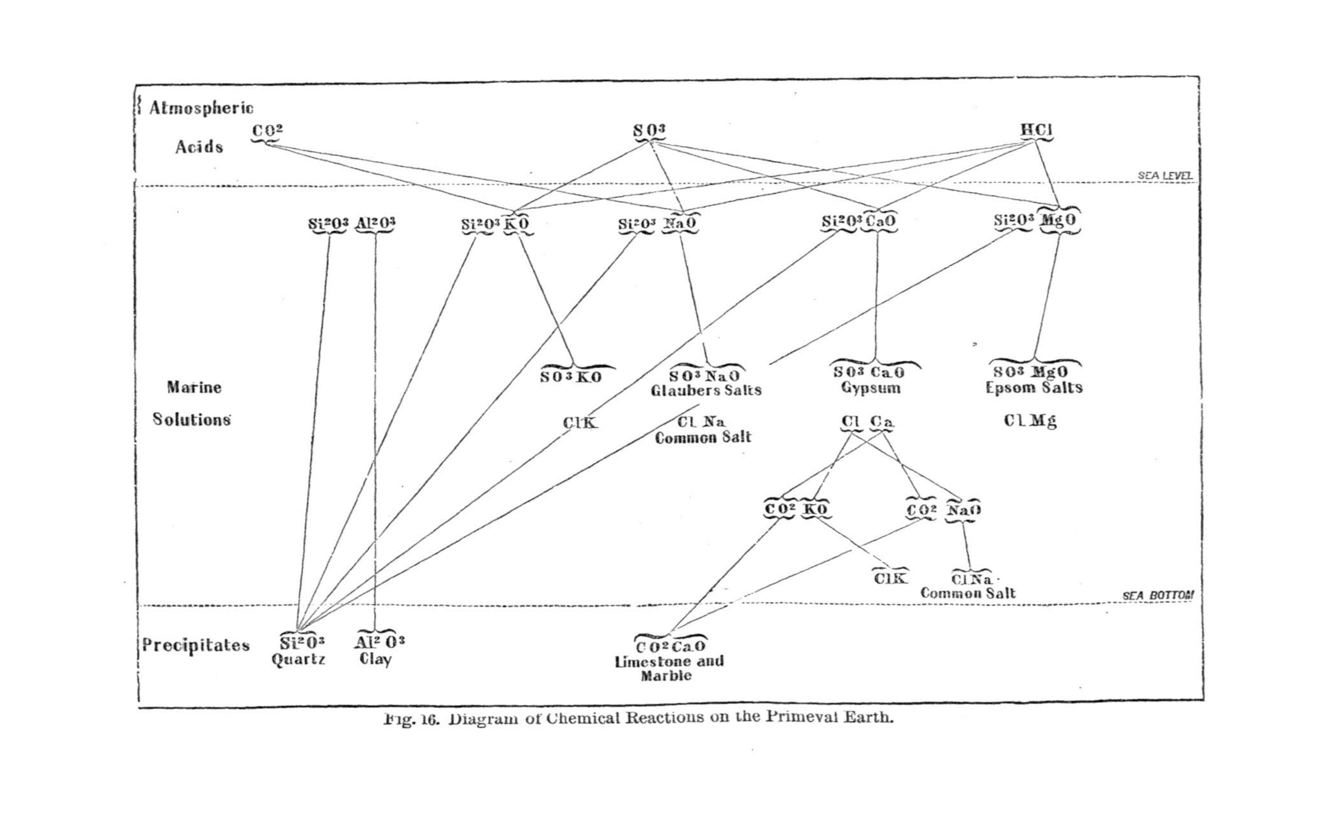

Fig. 16. Diagram of Chemical Reactions on the Primeval Earth.

will be rendered intelligible to all by the explanations in the text.

There seems to be but little poetry in the attempt to unravel the thread of chemical reactions which followed each other upon the earth in those dim and twilight ages; but it is certainly an inspiring development of late researches that the sceptre which chemistry sways over the modern world is the same which she wielded over the mute atoms of the forming crust.

It appears, from what has been suggested, that a portion of those ancient strata originated from sediments mechanically deposited, and another portion from chemical precipitates thrown down while the elements were adjusting themselves according to their strongest affinities.

The reader should not imagine that the proofs of these things are afar off. They lie within the scope of his own observation and verification. If you can not gaze upon the frowning summit of Katahdin, or the dark and lichen-covered sides of the Adirondacs, nor the upturned piles of stony lumber which make the ridges of the Appalachians, nor the acres of rocky floor torn up for your inspection along the shores of the upper lakes, examine some of the specimens which Nature has brought from those northern regions to your very doors. Scattered over your fields may be found fragments of the underlying unstratified granite and sienite, diorite and dolerite. Here, too, are fragments of rocks formed of the same constituents as these, but under a stratified arrangement. The most striking of these are the gneisses, where the various colored minerals set forth the stratification with distinctness. These came from the thick beds resting upon the crystalline foundation of the earth's crust. They are the ruins—a second time ruined—of some ancient rocky shore which the fury of the elements has reduced to sand. Here are boulders of quartz, liberated from its ancient combinations and precipitated in the bottom of the sea. Here are boulders of sandstone—vitreous, half-fused sandstone—better known as "hard-heads," which consist of grains of quartz produced by the grinding up of some more ancient quartz rock. These grains have been again cemented together, and a convulsion of Nature has sent them a second time vagrants over the surface of the earth. Here, too, are fragments of those ancient marbles, precipitated at the time when the partners of the ancient chlorides and carbonates formed new copartnerships for life. These all, rounded and battered by long travel, have come from their ancient homes in those northern regions where our continent first raised its head to scowl defiance at the supremacy of tempest and flood. They constitute, with numberless specimens of rocks of every other age, a grand museum, where every student of Nature may roam and study at his pleasure.

The chemical reactions, and precipitations, and sedimentary accumulations to which I have referred extended over an immense interval of time. During this long period materials accumulated at the bottom of the sea to the thickness of more than twenty-five thousand feet. Their geographical extent corresponded with that of the primeval sea. We find these rocks on every side of the globe, perforated here and there by the original granitic summits, which serve to point out to us the sites of the oldest islands. For our knowledge of the vast thickness of these older strata, their composition, and their wide American distribution, we are indebted to Sir William Logan and his associates of the Geological Commission of Canada. Sir William has ascertained that this stupendous pile of strata is properly divisible into two great systems, the lower of which he styles the "Laurentian," from the great river along whose valley they have been studied, while the upper is denominated the "Huronian," from the lake upon whose northern shores the upper members of the series are so finely exposed. The iron-bearing rocks of the northern peninsula of Michigan belong to the Huronian system, as well as those of Southeastern Missouri and Northern New York.

CHAPTER VII.

A RAY OF LIFE.

DURING the progress of that primeval age which witnessed the war of elements that I have already sketched, there was little opportunity for the unfolding of organic existence. The atmosphere was unfit for respiration; and the waters, if not too highly heated, were nevertheless charged with impurities destructive to both vegetable and animal life. It was a dreary and monotonous age, with nothing of that which now beautifies and diversifies the face of nature. The same sunlight fell upon the heaving waters of that tenantless and gloomy sea, and the same tide-wave performed its everlasting circuit round the globe. There was little diversity of weather or climate. The continents and mountain ridges, which give birth to oceanic and atmospheric currents, had not yet appeared above the wave. But there must have been a succession of seasons. The winter's sun, as now, went early to his couch, and his tardy rising belated the December mornings. His unequal favors to the different latitudes necessitated the trade winds and the great equalizing currents of the ocean. The higher density of the primeval atmosphere rendered it more retentive of the solar heat, and thus contributed greatly to diminish the rate of terrestrial cooling by radiation into space. Evaporation proceeded at a rapid rate, and condensation and precipitation were correspondingly copious. It was probably a stormy period, like the showery season which succeeds the protracted storm of the vernal equinox.

It would seem almost inevitable that the temperature and constitution of the primeval sea should be incompatible equally with vegetable and animal life. It is true that both plants and animals are now known to flourish under conditions of heat and cold, and chemistry, which are entirely at variance with the general notions of organic adaptability. Certain plants, for instance, are reported as flourishing in the boiling geysers of Iceland and the hot springs of California. Others make their habitat upon the snows of Greenland, and impart the ruddy glow of warmth even in the undisputed empire of frost. The germs of vegetable, and even of animal life, populate every element and every locality; and only a temperature of some hundreds of degrees suffices to rid a fluid exposed to the air of all the vitalized germs that inhabit it. The egg of an insect, stuck in the crevice of the bark of an apple-tree, endures the rigors of a Canadian winter; and the organized chrysalis seems, in many cases, to possess equal powers of resisting cold. It is unsafe, then, to attempt to determine at what epoch the waters of the primeval sea became sufficiently cooled and purified to receive the first organic forms. There was, in all probability, an earliest epoch that was completely destitute of organic forms. But, to ascertain its beginning and its end, Geology must yet apply herself to a closer study of the monuments of the gneissic age.

Reasoning deductively, it is equally presumable that vegetable life preceded animal life in order of appearance. Vegetable life is capable of enduring more extreme conditions. Vegetation could better tolerate the excess of carbonic acid in the atmosphere and the waters. Vegetation, moreover, is capable of drawing its sustenance from the mineral world, while animals rely exclusively upon organic food. The vegetable stands between the animal and the mineral, performing a sort of commissary function in behalf of the animal. The animal—even the carnivorous animal—implies the vegetable—requires the vegetable. All things considered, we are led to believe that plant life had a history upon our earth a full epoch before the existence of the lowest animals. There must have been a real Azoic Age. This deductive conclusion receives some support from inductive data. Petroleum, when existing in a state of wide or general distribution through a formation, is found to be traceable to vegetable organisms, generally marine plants, that have been reduced to a pulp and mingled with argillaceous mud before deposition. Petroleum is thus found in every formation, from the very latest down to the priměval gneiss. The actual presence of petroleum in gneissic strata affords a material prop to the doctrine of præzoic vegetation—a doctrine of no inconsiderable importance in establishing the harmony of the Mosaic and geologic records.

But a few months since geologists were equally ignorant of the existence of vegetable and animal remains through the entire series of Laurentian and Huronian strata, unless, perchance, the so-called "Cambrian" rocks of the Old World be of the same age as the Huronian—a conclusion which the eminent geologist, Dr. Bigsby, disinclines to accept. Geologists, it is true, drew the same inferences as now from the same data in reference to the existence of vegetal organization; but no actual or recognizable remains had been found, nor have they to this day. Greatly to the astonishment of the whole geological world, however, the abundant remains of animals have been discovered in strata which long antedate the most ancient in which a vegetable form has been descried. It was not by any means a rich fauna, but a single species, which populated the sea even in the Laurentian period. The faint tracery

Fig. 17. Structure of the oldest known Fossil (*Eozöon Canadense*). A thin section magnified. From a Photograph by Dr. Carpenter, London.

of its structure can be deciphered in thin, polished sections of Laurentian limestones and serpentines, when carefully examined under the microscope. These beings have been entombed in Canadian soil, and we have again to thank the energy and ability of the Canadian geologists for this modern revelation. The microscopic examinations have been chiefly made by Dr. Dawson, of Montreal, and have been fully corroborated by Dr. Carpenter, of London, England.

As might be expected, this being belongs in the very lowest rank of God's creatures. It is classed with the Foraminifera, in the group of Protozoa. It was related to the nummulite, whose skeletons have contributed so largely to the material of the Pyramids—monuments which perpetuate the memory equally of nummulites and Egyptian monarchs. It was related, also, to the little disc-like forms called *Orbitoides*, so abundant in the white limestone of the southern portion of the "Gulf States." Indeed, the kindred of this primeval forerunner of animal forms have been permitted to maintain existence in all seas, and in all ages, down to the present day. The type came upon the earth when nothing could dispute its pre-eminence. It has claimed a place among the ranks of higher animals in the ascending series, and does not shrink even from the face of man. Nay, the type maintains a foothold in the stagnant pools that gather upon the surface of the land, where man asserts peculiar supremacy. It demands our reverence for its antiquity. Let us pay it our respects.

Gazing through the microscope into a drop of water from some standing pool, our attention would scarcely be arrested by the sight of a little shapeless lump, which is as soft, and jelly-like, and inanimate, to all appearance, as any thing can be. But this is our Protozoan. It may be the species upon which science has imposed the name *Amœba princeps*, preserving the same gravity at the christening as if she had been naming a gorilla or a human animal, so impartial is science. We continue to gaze at our Amœba, and presently a little filament is extended like an arm, and perhaps immediately withdrawn. Soon a similar arm stretches forth in another position, and then another. Perhaps half a dozen arms extend themselves at once from dif-

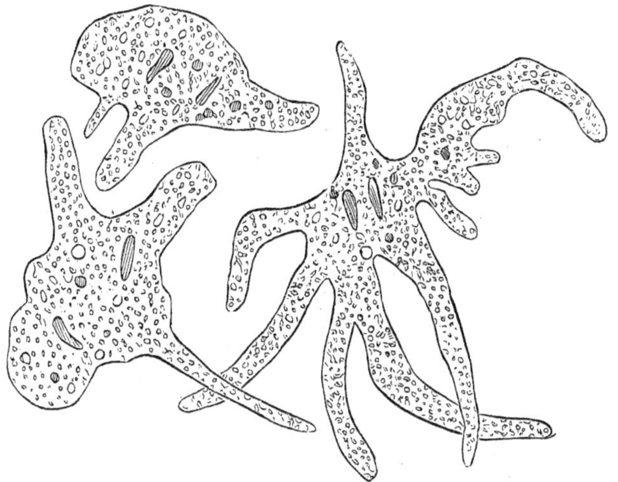

Fig. 18. Amœba Princeps, in different forms.

ferent sides, some long, some short, some thick, some thin. They shorten themselves, or entirely disappear, according to some inexplicable caprice. Now we discover the object of their movements. They are feeling for food; the Amœba is in search of his breakfast. As soon as a nutritive particle is touched, it is seized by one of these arms and introduced into the—mouth, do you say? No, indeed; the animal has no mouth. The food, however, gets inside by some means, and you may behold it there. It lies imbedded in the midst of the little lump of jelly. There is no stomach, no liver, no heart, no breathing organ, no head, no feet—in short, this animal is destitute of organs, except as it employs the whole body for every purpose. Whenever it seizes its food it extemporizes an arm for the work. Whenever it eats it must extemporize a mouth. Whenever it digests it must extemporize a stomach. It seizes, it eats, it digests, it breathes with the whole body. There are few animals, indeed, so utterly destitute of differentiation of the parts of the body. There is not the least division of labor. It symbolizes primeval society, in which every man does every thing that is done in the community.

Our Laurentian protozoan was as poorly furnished and as badly organized a being as this. But he possessed great advantages in point of size, and was, moreover, furnished with a stony armor—a wise provision, as one would think, for a creature that must buffet the storms which pulverized mountains, and defy the chemistry that dissolved granite. It only remains to effect the formal introduction to the reader. His name is *Eozoön Canadense*. [See Appendix, Note II.]

I said that the burial-place of this most venerable denizen of our planet was among the Laurentian rocks of Canada. Strange as it may appear, no vestige of animal organization has as yet been found among the overlying Huronian strata. It can not be doubted that life still continued upon the earth. It is possible that some of the most ancient forms of the Old World flourished during this age. Indeed, they herald the name of an Eozöon from Bohemia, and still another from the Emerald Isle. It seems certain that the latter had no contemporary and no rival for supremacy. He certainly was the first of the Fenians. But in America, so far as actual discovery goes, life touched the earth at a single point, and vanished again from view. This dawn of animal life was like the first gleam of sunlight that stole through a rift in the clouds of the primeval tempest, destined to be closed out again for a geologic age.

This enormous interval of time, down to the close of the Huronian, relieved of its absolute sterility of life by only a single species certainly known, we designate as Eozoic Time.

"The curtain falls, and the scene is changed." The crust, now becoming too large for the ever-shrinking nucleus, settles down to a closer fitting around it. The envelope, of course, must wrinkle, and the wrinkles must protrude their ridges, in some cases, above the waters. The horizontality of the primeval strata is thus broken. In some instances they are burst asunder, and the molten granite is poured out through the fissure. In other cases a huge back is simply elevated a moderate distance above the level of the sea. Weary of his old position, the giant, in adjusting himself in his new one, leaves his elbows protruding. Indeed, if we may extend this ugly figure, he may be represented as settling himself with an entire arm protruding above the waters which swept over North America. Beginning at the coast of Labrador, the arm—or ancient ridge of land—extends southwest to the north shore of Lake Huron. Here is the elbow. The fore-arm and hand extend thence northwesterly toward the Arctic Ocean. So it seems to be an arrangement of Nature that "Johnny Bull" shall continue to thrust his elbows into the sides of Young America! We acquiesce, for the present, in this arrangement. Meanwhile, other spirits will be summoned from the "vasty deep," and teeming life will appear upon the stage in the next act of the drama.

CHAPTER VIII.

THE FRONT OF THE PROCESSION OF LIFE.

THE spirits have come forth. The life-giving *afflatus* has been breathed into multitudes of organic forms which now teem in the PALEOZOIC sea.

"Say, mysterious Earth! oh say, great Mother and Goddess!
Was it not well with thee then, when first thy lap was ungirdled,
Thy lap to the genial Heaven, the day that he wooed thee and won thee?
Fair was thy blush, the fairest and first of the blushes of morning;
Deep was the shudder, oh Earth! the throe of thy self-retention;
July thou strovest to flee, and didst seek thyself at thy centre!
Mightier far was the joy of thy sudden resilience; and forthwith
Myriad myriads of lives teemed forth from the mighty embracement,
Thousand fold tribes of dwellers, impelled by thousand fold instincts,
Filled as a dream the wide waters."

The long period of almost total lifelessness—the EOZOIC TIME—it will be remembered, was brought to a close by the upheaval of a long ridge of land, extending from the coast of Labrador to the northern shores of the great lakes, and thence northwest to the Arctic Sea. Corresponding upheavals took place on other continents. A convulsion could not jar one half the globe without being felt upon the other half, and hence it is that all the grand revolutions of geology were simultaneous, and the histories of different continents are divided into corresponding chapters. We confine our attention, however, to North America. The germinal ridge consists of an axis or nucleus of granitic material, and on each side of a series of gneissoid and other eozoic strata sloping like the roof of a house from the central and highest part. We know that this upheaval took place *after* the deposition of the eozoic strata, because those strata could not have been deposited in their present tilted position (Fig. 19). We know that it took place *before* the deposition of the next series of strata, because

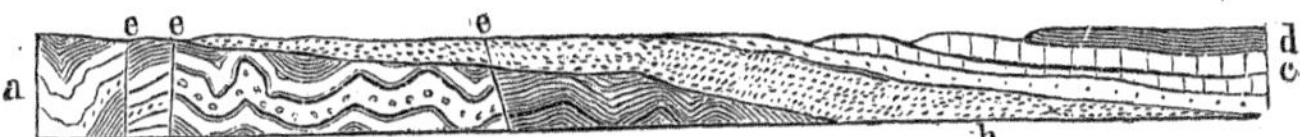

Fig. 19. Disturbed condition of Eozoic Strata.
a. Eozoic strata, tilted and contorted by disturbances occurring before the deposition of the Paleozoic strata, b, c, d. e, e, e. Faults or dislocations of Eozoic and Paleozoic strata, evidently of later date than the contortions of a.

these strata were not tilted by the upheaval, but continue to present their horizontal edges against the inclined faces of the eozoic beds. Thus the precise relative period of this upheaval is fixed.

Consider the geography of the North American continent at this date. An angular ridge of land (Fig. 20) is all that

Fig. 20. The Germ of the North American Continent.
a, a. The two branches of the continent. b, b. Islands. (The modern continent is indicated by dotted lines, the rivers by broken lines.)

then existed. The Rocky Mountains and the Alleghanies were not yet born from the deep. Where were the United States? Where the broad valley of the Mississippi, and the wide-extended plains of the far West? Beneath the wave, and receiving the sediments of the same sea which rolled over the future sites of Babylon, and Tyre, and Athens, and the seven hills of the "Eternal City." The generations of men yet slumbered in the chambers of futurity. The order of Providence had assigned them their position in the grand procession of life which was now beginning to move, and the scouts of which had passed by in the preceding age; but we must wait for man till a long line of grotesque and marvelous forms has marched before our view.

The van of this procession was led by some of the humbler forms of God's creation. We shall indeed look in vain for a type of existence of simpler mould than the Laurentian Eozoön. It is likely that beings akin to this accompanied the shoals of higher forms which sprang into existence at the morning dawn of the Silurian Age. But if they lived, the record of their existence has been effaced from the earth. Like the deeds and the sufferings of the men who kept company with the extinct quadrupeds of Europe, and chased the fur-clad mammoth across the steppes of Siberia, their very existence is reached only by conjecture, and the activities which made up life with them have all been locked up with the arcana of the past. The creatures whose relics we have disentombed were more highly gifted than the Eozoön, and were launched into being under a great variety of forms. The oldest Silurian rocks of North America are perhaps those revealed to science upon the island of Newfoundland by the assiduity of the Canadian geologists. Their records have been recently studied by the paleontologist Billings, of Montreal, an investigator eminent for acuteness and for the importance of his paleontological discoveries. An assemblage of strata named by him the "St. John's Group" is described as underlying rocks that had heretofore been regarded as forming the very base of the Silurian system in America. These St. John's strata may probably be regarded as inclosing the remains of the first considerable fauna that ever lived within the limits of America. Our knowledge of these primeval relics is as yet very imperfect*, being limited to one crinoid, two brachiopods, and half a dozen genera of trilobites. Though they mark generally a great simplicity of organization, one can not but be astonished that in the very outset of animalization upon our globe so high a rank and so great variety of types should have been manifested. If we are to judge from that which is known rather than that which is conjectured, we are compelled to conclude that the varied forms of animal life did not come into being by a gradual evolution from the Eozoön, but as so many original utterances of the all-skilled Artificer of creation.

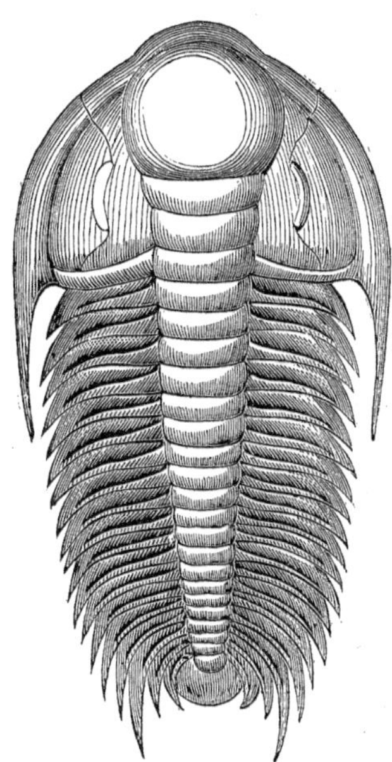

Fig. 21. Paradoxides Harlani (×⅙). St. John's Trilobite.

Of the "Potsdam group" of strata [see Appendix, Note III.], and the organic remains which they inclose, we have learned somewhat more. The "Potsdam sandstone" at the

* Billings (E.): "Catalogues of the Silurian Fossils of the Island of Anticosti," p. 79.

bottom of the group, was long regarded as the oldest fossiliferous rock in America. It is certainly not far from the lowest horizon of the primeval cemeteries which hold the dust of the first denizens of our planet. This sandstone is sometimes whitish or grayish, but often of a dull red color, and sometimes slaty; and except within the area covered by the St. John's group, it is found resting upon the upturned edges of the Eozoic strata. Observe that it is a sandstone. Now we know that in the waters of the present day, sands are accumulated only in comparatively turbulent and shallow regions. In calmer and deeper waters, the sediments are necessarily finer, as only the finest particles can be transported by the slowest moving currents (compare Fig. 15). Moreover, many a layer of this ancient sandstone, when uncovered to the light, presents us with veritable ripple marks—such as the waves are making to-day in the fine sand of the shallow water near the beach—sand-ripples which have been preserved unmarred for millions of years, and unite with other proofs that the bottom of the Protozoic sea was not beneath the reach of the agitations of its surface. This interesting sandstone was first attentively studied at Potsdam and Keeseville, in

Fig. 22. Cliffs of Potsdam Sandstone on the Au Sable River, New York.

Northern New York, and the geologists of the Natural History Survey of that state christened it accordingly from one of these localities (Fig. 22). It was burst through at some subsequent period by some of the granitic rocks now constituting the region of the Adirondacs—or, if some of their domes were already islands in the Eozoic ocean (b, Fig. 20), their massive walls have been heaved to higher altitudes by later efforts of nature, since the rocky wrappings of their flanks have been raised to inclinations which prove disturbance subsequent to their deposition. This sandstone extends southward into Pennsylvania, where, at a still later period, it was upheaved by the convulsion which brought the Alleghanies to light. Still farther south, in Virginia, Tennessee, and Alabama, this ancient sea-bottom has been brought up at intervals along the dislocations of the Appalachian range; while on the west of the Mississippi it comes up again in the highlands of Texas and Arkansas, in Eastern Missouri, and the Northwestern States, and has been broken through in the Black Hills of Dacotah by a comparatively recent protrusion of granite. From Northern New York it trends down the valley of the St. Lawrence, while in the opposite direction it crosses over to the northern shore of Lakes Huron and Superior—underlies the western portion of Superior, and spreads itself out over vast areas in Wisconsin and Minnesota, whence its main outcrop sets out for the region of McKenzie's River, on the arctic slope of the continent.

These localities and regions are but the present places of outcrop or exposure of a solidified bed of sands, which was accumulating in the bottom of the ocean at the time of which we speak. What of the beings that enjoyed the throb of life in those ancient waters? The renowned paleontologist of New York, Professor James Hall, has made us acquainted with but two or three distinct creatures from the whole extent of the typical region of the Potsdam sandstone. These have been named *Lingula prima* and *Lingula antiqua*. They are little bivalve shells belonging at the bottom of the class Brachiopoda, which is nearly the lowest class among molluscs. As destitute of the senses as an oyster, they were equally incapable of locomotion, being anchored to the bottom by a fleshy stem or peduncle which issued through the hinder part of the shell, and had an internal organization which was even more rudimentary and homogeneous than that of the "bivalve," which has become the type of insensibility and stupidity. The same little shells have been observed in Northern Michigan, in Minnesota, in Wisconsin, in Alabama, and even in the Old World, every where occupying a position in strata which were accumulated at the same time as the Potsdam sandstone. In many instances the extent to which the number of individuals was multiplied is truly amazing, while the whole catalogue of species of molluscs in this sandstone scarcely reaches half a dozen. With these bivalves, in Wisconsin and Minnesota, are associated incredible numbers of *trilobites*. As might be expected in deposits formed under such conditions as gave rise to sands, the trilobites are found generally in a greatly damaged condition. These Northwestern cemeteries have been mainly explored by Dr. D. D. Owen, Professor James Hall, and Dr. B. F. Shumard. The writer has also had the opportunity to bring

Fig. 23. Lingula prima.

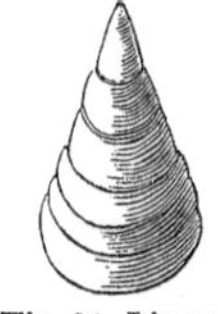

Fig. 24. Lingula antiqua.

Fig. 25. Modern Lingulas anchored to a support.

to light some hitherto unrecognized forms. Still other species have been made known from Texas by Dr. Shumard and Dr. F. Römer. Trilobites belong to the lower—not the lowest—part of the sub-kingdom of Articulates. The radiates and the great mass of molluscs hold inferior rank, and yet throughout the world we find these lower strata characterized by a profuseness and variety of trilobite remains which are not approached by the molluscs or the radiates. Many investigators have contributed to our knowledge of these primordial creatures, but to none are we so deeply indebted as to M. J. Barrande, who has enriched with marvelous details his great work upon the "Silurian System of Bohemia." He has traced them through the various stages of their embryonic development, and shown that they underwent metamorphoses to some extent similar to certain insects. Varying in size from a pea to a foot or more in length, they had the jointed external shell of a lobster, and could roll themselves together like a hedgehog for the purpose of passive protection. Multitudes of them are found folded in this condition (Fig. 27), intelligible witnesses of an instinctive shrinking from the death-pang, which, even in this early age, was the means employed by Providence to secure the lives of his sensitive creatures. With all except the lower forms the eyes are distinctly discernible, and even in these the places for the eyes are visible, and there is no reason to suppose they were blind. In the others the eyes are curiously compound, like those of the common house-fly. Did the reader ever examine the eyes of the domestic fly with

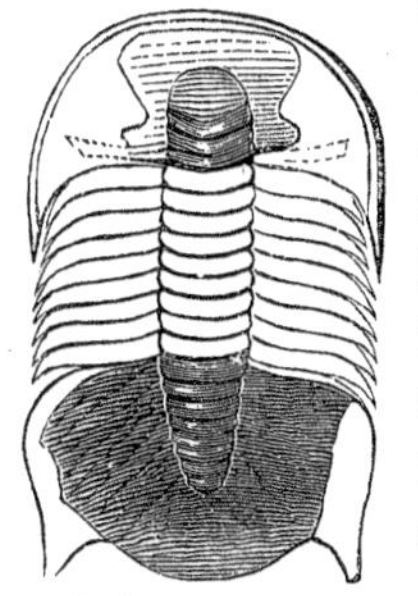
Fig. 26. Dicellocephalus Minnesotensis.

Fig. 27. Side view of a Trilobite (*Calymene senaria*) rolled up.

a hand magnifier? If not, the beautiful and perfect structure which it displays will compensate for the trouble of procuring the means to make the observation. Some scores of little lenses, arranged with the most perfect symmetry, each set in its little telescopic tube, form upon the retina of the little insect the various portions of an image of some external object. Such eyes had the trilobite (Fig. 28). It is marvelous that such delicate structures have been so

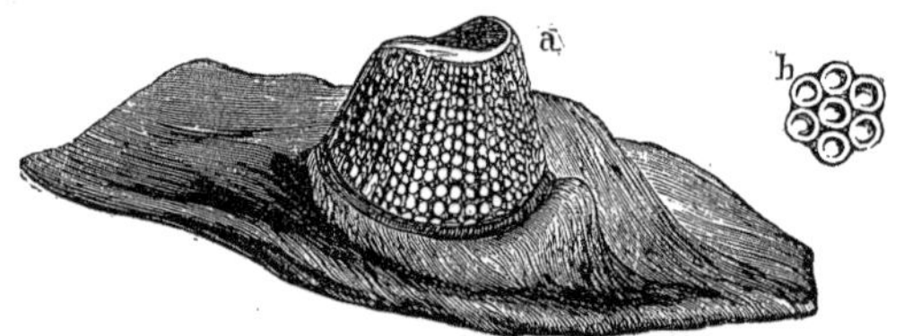

Fig. 28. The eye of a Trilobite magnified.
a. Eye of *Asaphus caudatus*. b. A few facets of the eye of *Calymene macrophthalmus*.

perfectly preserved as in some trilobites which I have examined from the neighborhood of Dubuque, Iowa. These, however, existed in the period following the Potsdam. The trilobite was tri-lobed in two respects. Longitudinally, the oval form was divided into head, body, and tail; while in the other direction a couple of lengthwise grooves divided the animal into middle, right, and left lobes, or regions.

In this earliest scene of animalization, mollusks and radiates play comparatively an inconspicuous *rôle*. But it must be remarked that both these types of existence had been introduced. Among the molluscs we have found, besides the representatives of Brachiopods already mentioned, a few other members of the same class, and also some coiled univalves, which belong to the higher class of Gasteropods. Among the radiates we have in the Old World a few representatives from the middle of the sub-kingdom in point of rank, while among Protozoa we find a few forms related to sponges, with calcareous instead of horny skeletons. In the epoch immediately following this, animal life rose to a slightly higher grade, and unfolded in a great variety of subordinate types. Before the close of the Potsdam period—before the deposition of the sediments which formed the limestone and marls of Cincinnati, and have given character to the far-famed "blue-grass region" of Kentucky—life had been ushered upon our globe in such richness and variety, that not only had three of the four fundamental plans of animal organization been realized, but all or nearly all the various classes of the three lower sub-kingdoms had been fairly represented.

Many extensive regions of the Potsdam sandstone and overlying calciferous sand-rock are, nevertheless, almost, if not quite destitute of the traces of organic existence. Along the south shore of Lake Superior is a sandstone once regarded as belonging to the Potsdam, but probably, in part, of the age of the "Calciferous," in which we search in vain for any of those fossil remains so common in Minnesota. We find nothing but the imprints of soft sea-weeds (Fig. 29)—things like films of jelly, which have left their imprints upon the coarse rock, and have transmitted to us a knowledge of their existence and nature, while the traces of an army's march are obliterated by the vicissitudes of a single season.

The Lake Superior sandstone, whatever its geological age, is a formation of remarkable interest, both in its relations to the basin of the largest lake in the world, in its relations to the world-renowned copper deposits of the region, and, not less, in its relations to some of the finest scenery of the continent. The remarkable interest of this formation was first pointed out by Dr. Douglass Houghton—a name more honored and beloved among the Wolverines than any other in the lists of science.

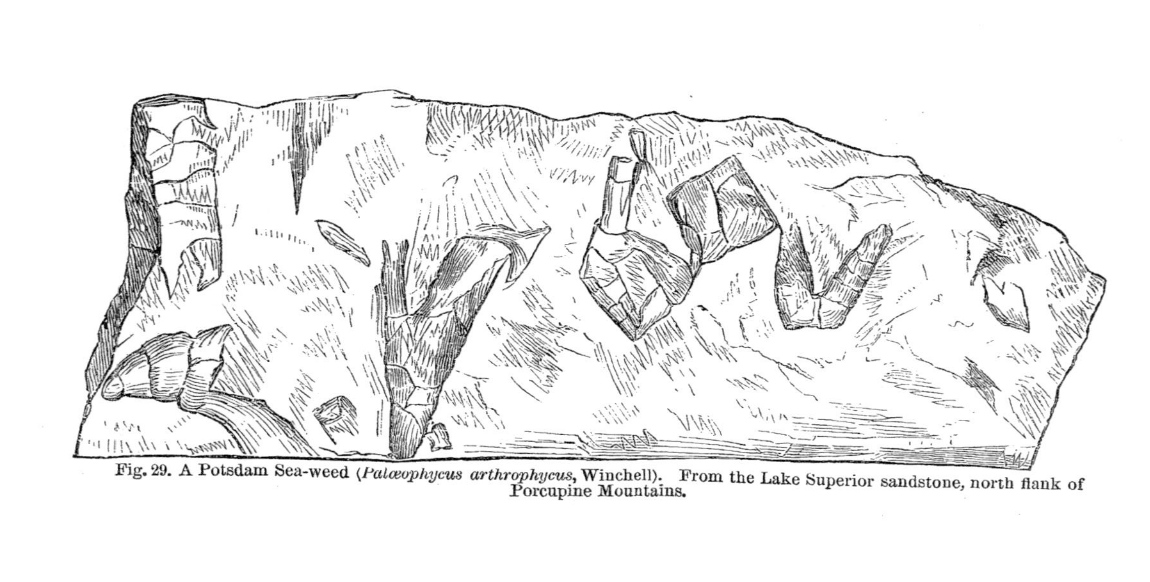

Fig. 29. A Potsdam Sea-weed (*Palæophycus arthrophycus*, Winchell). From the Lake Superior sandstone, north flank of Porcupine Mountains.

The convulsions which closed the Potsdam period protruded through this formation enormous vertical walls of molten rock, known by the general name of "trap," or dolerite, or delessite. The bursting through of these igneous materials tilted up the broken edges of the sandstone, and formed between the lines of outburst deep valleys, which have become the bed of the lake. The sandstone which plunges beneath the water's surface on the northwest side of Kewenaw Point reappears on Isle Royale, which was formed, like Kewenaw Point, by an outburst of dolerite (Fig. 32).

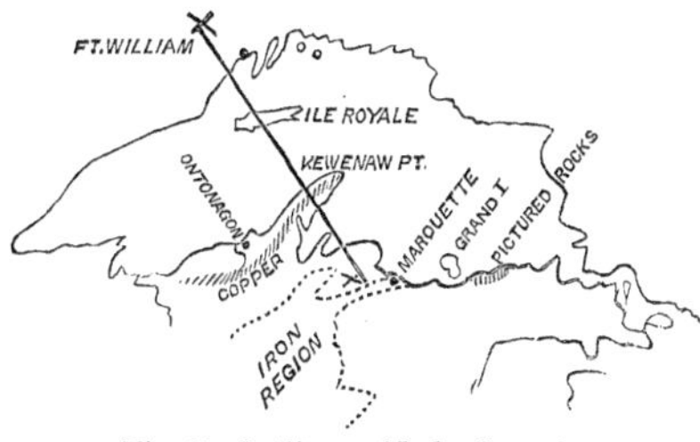

Fig. 30. Outlines of Lake Superior.

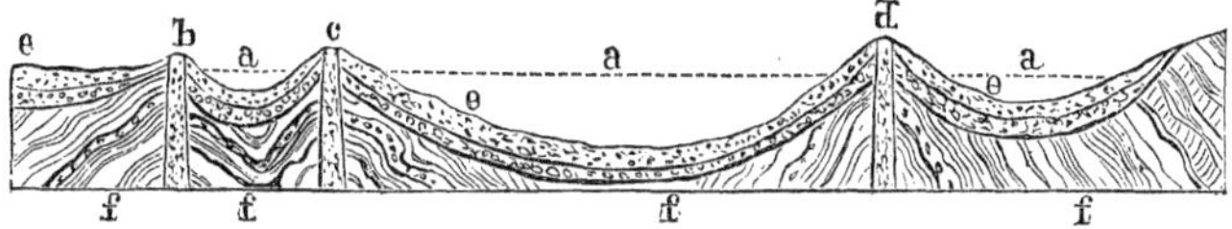
Fig. 31. Section across Lake Superior, along the line XX, Fig. 30.
a, a. The water level. b. Trap outburst north of the lake. c. Trap outburst forming Isle Royale. d. Trap outburst forming Kewenaw Point. e. Lake Superior sandstone and conglomerate. f. Eözoic and other rocks underlying the sandstone.

From the north side of Isle Royale the sandstone glides under the water again, and reappears upon the northern shore of the lake. The basin of the lake is therefore a geological valley—a "synclinal" valley—formed by the igneous eruptions upon the northern and southern shores. Its origin, as will be seen, is entirely different from the origin of any of the other lake basins of the chain.

The escape of the molten rocks of the region fused out the copper and silver, which were disseminated through the neighboring strata, and accumulated them in masses of great commercial importance. An enormous dike of dolerite, three or four miles in width, runs like a backbone from the head of Kewenaw Point, southwest and west, to the neighborhood of Ontonagon (see Fig. 30). Upon this the copper mines of the region are located. In each direction from this backbone slopes the sandstone which was upturned by it (Fig. 31). Directly across this adamantine ridge passes a stupendous cleft, which has been filled with water from the lake. Thus has been formed Portage Lake, a narrow, winding body of water, which vessels navigate from the east to within two or three miles of Lake Superior on the west. Commerce has undertaken to complete the work begun by Nature, and soon Kewenaw Point will be an island.

Toward the eastern extremity of Lake Superior the formation reposes in nearly horizontal beds, and the erosion of the lake along the southern shore has carved out bold escarpments which arrest the attention of every traveler. These have been named the "Pictured Rocks" (Fig. 32), from the diversified colorings of the various portions of the sandstone mass. It is a dangerous coast, and no species of craft ventures within inspection distance except in calm and settled weather. The high bold wall looks sheer across the lake, and the storm-wind rolls tremendous surges against its stern, defiant face. They have excavated caverns which a canoe may traverse, and in which the imprisoned billows howl with the resonance of a Nahant purgatory. They have carved out mimic architecture and spread a mimic sail. In times long gone by they have flanked a stubborn bluff, and cut it off from the main land by a military "ditch" of the most successful kind, since it is a mile wide, and is navigated by the largest vessels. The isolated piece is known as Grand Island.

"The range of cliffs to which the name of Pictured Rocks has been given," say Foster and Whitney, "may be regard-

Fig. 32. The Miners' Castle—"Pictured Rocks," Lake Superior.

ed as among the most striking and beautiful features of the scenery of the Northwest, and are well worthy of the attention of the artist, of the lover of the grand and beautiful, and of the observer of geological phenomena."*

The first display of architectural mimicry witnessed in coasting eastward is a salient mass of sandstone known as the "Miner's Castle," presenting the turreted elevation and arched and massive doorway of some ancient feudal seat. The height of the doorway is about seventy feet, while the tops of the turrets are one hundred and forty feet above the lake (Fig. 32).

About five miles farther eastward the cliffs attain an elevation of about one hundred and seventy-three feet, presenting a series of sinuosities or scollops hewn out by the action of the waves. One of the grandest and most regular of these was named "The Amphitheatre" by Foster and Whitney. Still farther eastward this scolloped contour graduates into extravagant carvings, which have wrought the mural wall into wierd Titanic mimicry of architectural forms.. Vast tablets from the upper courses of the wall, sapped by the agency of eroding waves, have tumbled down and strewn the beach in places with fragments which lead the traveler to believe he is clambering among the ruins of gigantic temples shattered by an earthquake shudder. A group of these fallen fragments presents a striking similitude to the jib and mainsail of a sloop full spread,

* Report on the Geology of the Lake Superior Land District, part ii.; p. 124 (1857). These authors have given the fullest and most exact account of the Pictured Rocks that has yet been published. Schoolcraft, at an early period, undertook to describe this range of cliffs and illustrate the scenery, but with very poor success. Harper's Magazine, vol. xxxiv., p. 681 (May, 1867), contains a lively and interesting paper on the "Pictured Rocks," embodying several good illustrations. Some of the following views are from photographs by Watson, taken on an excursion by a party from the University of Michigan in 1868, under the leadership of Dr. A. E. Foote.

Fig. 33. The Sail Rock, at the "Pictured Rocks," Lake Superior.

Fig. 34. The Grand Portal—"Pictured Rocks."

and hence has been dubbed by the *voyageurs* "the Sail Rock" (Fig. 33).

A mile farther east we reach "The Grand Portal" (Fig. 34). This is an enormous arched gateway one hundred feet in height and one hundred and sixty feet broad, opening into a magnificent vaulted passage some three hundred feet deep, and expanding into a massive dome. These apartments, with their ramifications, have been hewn in an enormous quadrangular block of brown sandstone projecting sheer into the lake six hundred feet, and presenting a front of three or four hundred feet, with a frowning façade lifted full one hundred and thirty-three feet above the

Fig. 35. Camp on the Beach near the Chapel.

water which bathes the foundations and resounds through the vaulted passages of this most magnificent of Nature's cromlechs.

The last and most grotesque of these mural structures is "The Chapel." At the height of forty feet above the lake

Fig. 36. The Chapel—"Pictured Rocks."

is a rocky floor, from the four angles of which rise curiously wrought columns of masonry in thin and regular courses. These support a massive vaulted roof that covers a rustic auditorium forty feet in diameter and forty feet high, which suggested the name of the structure. At the base of one of the columns is excavated an arched niche, which may be reached by a flight of steps formed of the retreating layers of the sandstone. This is the pulpit. In front lies a tabular mass of rock which answers for the desk, while an isolated block on the right represents an altar. "If the whole had been adapted expressly for a place of worship, and fashioned by the hand of man, it could hardly have been arranged more appropriately. It is hardly possible to describe the singular and unique effect of this extraordinary structure. It is truly a temple of Nature—'a house not made with hands.'"

Hard by this chapel, erected by the hand of Nature to symbolize the devotion which Nature's solitudes inspire, is one of Nature's preachers—a beautiful cascade—lifting up its voice perpetually in hearing of the spirits of the primeval wilderness in the rear,

Fig. 37. Chapel Falls—"Pictured Rocks."

> "Mingling its echoes with the eagle's cry,
> And with the sounding lake, and with the moaning sky."

Not far east from here rises a stupendous dune of sand, or, rather, a promontory of uncemented sand and clay, capped by a shifting dune. The grinding action of the waves has pulverized a cubic mile of sandstone and superincumbent drift, which has been strewn over the lake's bottom. The nervous wind-gust has wrested it from the water, and made it a plaything of its own. Dried by the sun and air, it has been driven inland till the forest is submerged, and a shining promontory called Grand Sable lifts its forehead four hundred feet above the lake—a landmark for the mariner and a marvel to the lover of Nature.

CHAPTER IX.

DISCOVERY OF THE PROGRAMME.

THE reader will have observed that the primal or Potsdam sandstone has been traced all around the circuit of the central United States. There is no doubt that it underlies all the region embraced within the circumference of its outcrop. Indeed, at Columbus, Ohio, an artesian boring has probed the crust 2000 feet and more, and found this sandstone in its proper place. At Lafayette, in Indiana, Louisville, in Kentucky, St. Louis, in Missouri, and Chicago, in Illinois, similar deep borings have been executed, and the succession of strata, as far as the borings extend, has been exactly such as geology expected; and there can not be a doubt, that wherever exploration should be made throughout the wide extent of the area indicated, the Potsdam sandstone would be found occupying its proper position at the bottom of the Silurian series of strata. As the same formation has been upheaved, at intervals, along the whole distance to the Rocky Mountains, geologists have arrived at the conclusion that the entire area of the United States and Territories was, during the Lower Silurian Age, the bed of a comparatively shallow sea.

This conclusion leads to a generalization of the highest interest. How came the central area of the North American continent a basin of shallow water? We can only infer that, at this early period, the Alleghanies on the east, and the Rocky Mountains on the west, had already begun to be lifted above the general floor of the ocean. The United States were an immense continental lagoon—a submarine plateau, such as now exists in the North Atlantic, upon which the telegraphic cable has been laid. The outline of the continent was consequently marked out while yet in embryo. The foundation of the Alleghanies was laid ages before the superstructure rose above the waves, and exposed to the light of day the predestined trend of the Atlantic coast of our country. But we trace the development of this idea back to a still remoter period. Note the trends of the primeval ridge (Figs. 19 and 39) which still lies thrusting its angle down into the northern notch of the "great lakes." "N o r t h e a s t and northwest" was the language of that early-uttered decree which foredetermined the shape of the continent which was destined to become the "land of the free." That primal ridge was its earliest germ. Successive annexations to this germinal continent have been uniformly toward the southeast and southwest. This primitive ridge was not alone an early prophecy of the trends of our present coast

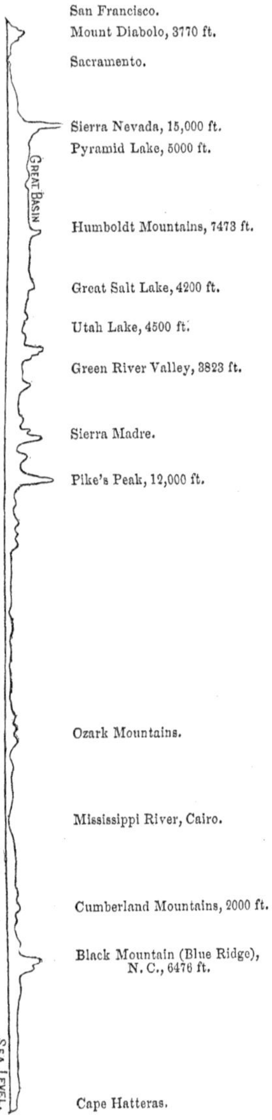

San Francisco.
Mount Diabolo, 3770 ft.
Sacramento.
Sierra Nevada, 15,000 ft.
Pyramid Lake, 5000 ft.
GREAT BASIN
Humboldt Mountains, 7473 ft.
Great Salt Lake, 4200 ft.
Utah Lake, 4500 ft.
Green River Valley, 3823 ft.
Sierra Madre.
Pike's Peak, 12,000 ft.
Ozark Mountains.
Mississippi River, Cairo.
Cumberland Mountains, 2000 ft.
Black Mountain (Blue Ridge), N. C., 6476 ft.
SEA LEVEL.
Cape Hatteras.

Fig. 38. Section across the Continent of North America.

lines. In its upper angle lies Hudson's Bay, whose place was designated as soon as it became the bottom of a submarine valley. The southern slope of the ridge became the water-shed which was to supply the great lakes and the St. Lawrence. The St. Lawrence finds its outlet to the ocean in a valley parallel with the ancient ridge. The peculiar notch from Georgian Bay to the head of Lake Erie, and thence to the Niagara River, is conformed to the sal-

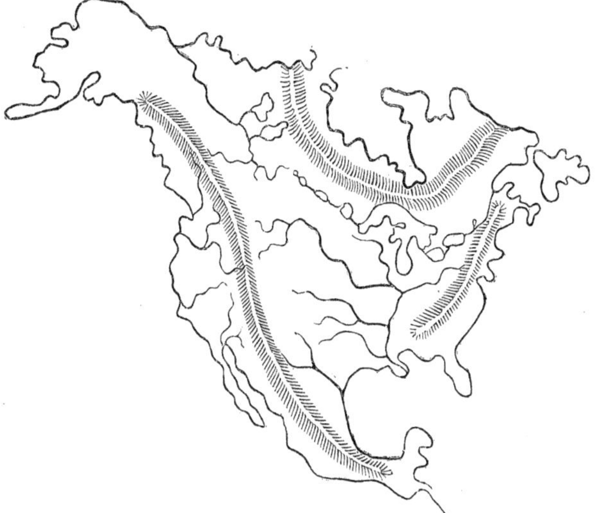

Fig. 39. Hydrographic and Orographic Outlines of North America.

ient angle of the same ridge. The "great lakes" themselves are but links in the vast chain of lakes extending to the Northern Ocean, accumulated in a valley inclosed by the western branch of the continental nucleus on the one hand, and the occidental ridge of the continent on the other. The Mississippi pursues its course along the bottom of the depression between the Appalachian and Rocky Mountain ridges, while the McKenzie—the Mississippi of the North—is its counterpart, holding possession of the northward prolongation of the same depression. Or, to present the generalization in another form, the primordial ridge, with its northeast and northwest branches, holds Hudson's Bay in its embrace. The Appalachians and the Rocky Mountains constitute the two branches of a secondary ridge, which do not meet toward the south. One of these branches points toward the prolongation of Florida and the peninsula of Yucatan, and the other toward the prolongation of Mexico and Central America, with the Gulf of Mexico—the Hudson's Bay of the South—occupying a depression between them. The space between the primary and secondary ridges has two systems of drainage—one toward the north, and one toward the south. Each system has two branches. In the northern system the branches diverge from the lake region toward the northeast through the St. Lawrence, and the northwest through the McKenzie. In the southern system the branches converge through the Ohio and the Missouri, and discharge themselves by one outlet through the Mississippi. Thus the whole hydrographic and orographic system of North America has been determined by the location of these skeleton ridges—pieces of the framework which, though for unnumbered ages they were yet unborn from the deep, were nevertheless working out the configuration and the topography of a continent. Indeed, as the secondary pair of ridges was but a reduplication of the first, or Laurentian pair, we find that the innumerable hydrographical and topographical features of our continent have taken their point of departure from the Laurentian ridge as an initial and germinal area. Finally, the trend and conformation of our eastern coast are what has turned our "Gulf Stream" to the northern shores of Europe, to mitigate the climate of a little inhospitable island in the latitude of

Labrador, and nurture its people to become the "guardians of civilization."

These comprehensive views of continental development have been especially wrought out by that Christian gentleman and unrivaled scholar, Professor J. D. Dana, of New Haven.

It is wonderful to behold one of Nature's great plots worked out with such undeviating unity of purpose. Though incalculable ages have elapsed since the nucleus of the American continent was lifted above the waves, we find the announcement then made to have been faithfully prosecuted to the end. What convincing proofs of the unity of the Creative Intelligence! The plastic rocks have always been moulded by the hands of the same all-providing Artificer. How it exalts our apprehension of his infinite attributes to behold him bringing into existence a series of secondary causes, so simple in themselves, but working out a succession of results so complex in their details, and presenting a history stamped with such uniformity of plan, such harmony of parts, and such wisdom of design! But these are only his doings in the material world. When we contemplate the manifestation of his attributes presented to us by animated nature, every one imbued with the spirit and love of truth is compelled, with the poet, to exclaim,

"An undevout *philosopher* is mad."

We turn, then, to consider the method which reigns among creatures exalted with the gift of life.

Who has not been amazed at the endless variety of animal forms existing upon the earth? There seems to be no conceivable conformation, no possible situation, no circumstances of element, climate, food, or condition, that have not been made the fitting and essential conditions of some type of conscious existence. One animal dwells on the land, another in the soil, a third in the air, a fourth in salt water, a fifth in fresh; one burrows in a log, another in a rock, a third in the mud, a fourth in the flesh, or brain, or liver, or even the eye of another animal. Ponderous quadrupeds move through the jungle, wily serpents glide among the reeds, the centipede crouches under a stone, the minnow darts beneath the sedgy bank, and the lazy oyster sleeps in the mud at the bottom of the bay. We place beneath the microscope a specimen of the mud in which the oyster spends his drowsy life, or even a sample of the water in which the familiar frog delights, and lo! another world is revealed to our vision—vegetal and animal life in forms as varied as all that the unassisted eye has seen in the greater world.

Nor is this all. Every one has read of forms long since extinct—of strange and monstrous forms that sported upon the earth before the empires of the brute creation had been subjugated by the intellect of man. A stone-mason of Cromarty has introduced to the world the *Asterolepis* of Stromness, and the *Cephalaspis* and *Pterichthys* of the "old red sandstone"—fishes which the most learned had at one time almost decided to throw into the company of turtles. Mantell has amazed us with stories of the *Iguanodon*, an immense lizard, believed by him to have been sixty feet in length, which crawled over the slime of the latest part of the Jurassic period. These all were forms of the middle ages of the world's history. As we run back through the æons preceding, we tread upon the graves of myriads of beings which in their day swarmed in the depths of the sea, but whose lineage and likeness are now known only in history. We push back through the dim dawn of being, and stand upon the sandy shore of that uneasy sea in which Creative Power first essayed to mould the plastic clay into animal forms, and plant in them ethereal fire. How reverently do we turn up the cleaving stone, and gaze upon a little *coral*, a *Lingula*, or a *trilobite*, and think that these were the forms which God first exerted his skill upon, and placed first in possession of our round and verdant planet! And how different those beings from all we know upon the earth to-day! What an infinite range of aptitudes between that humble *Lingula* and the majestic mien of man! Such is the exhaustless fertility of God's conception.

We place ourselves, then, upon the threshold of animal existence, and inquire what course creative Power will pursue. Shall we witness a series of experiments for the slow perfection of a plan—models and methods tried and abandoned—detached essays, having no intelligent connection with an ultimate or central scheme? With a finite intelligence such experiments would have been unavoidable; but Nature has served no apprenticeships; the end has been contemplated from the beginning.

There are two things which strike the attention of every one who studies the history of the ancient populations of our globe. First, their forms and features, their habits, and the details of their living, are often in wide contrast with any thing we behold at the present day. Secondly, while so peculiar in their *details*, their *fundamental* features are *identical* with those of existing animals, so that we call them by the same generic titles—corals, shells, crustaceans. And if we scan the long line of being from the Laurentian to the present, we shall find nothing which may not be embraced under the most general designations which we apply to existing animals.

Now which of these two features of the fossil world is most instructive? Their wild and extravagant forms astonish us, and attract the curiosity of the marvel-loving public. Their identity of fundamental plan impresses us with awe and reverence, and breathes the thoughts of a world-embracing scope of intelligence. The first converts the animal creation into a vast menagerie for the curious to wonder at; the latter shows it to be a lesson of wisdom traced by the finger of the Omniscient himself.

Let us see what is the nature of this identity of plan which runs through all existence and all time. It is a wonderful fact in Nature. From the epoch of the St. John's molluscs and the Potsdam trilobites, through all the dreary ages of the earth's preparation for man, but four fundamental types of animal structure have ever existed. All the varied forms of extinct monsters have been constructed upon one or the other of these four fundamental plans. Throughout the wide range of existing beings—inhabiting the deep sea, populating the air, swarming over the land, and the forest, and the jungle—countless equally in the number of individuals and in the number of distinguishable species—we discern but the same four foundation plans of structure which we find exemplified in the creations of the ancient world. As the seven fundamental intervals of the gamut have in their endless combinations afforded us all the varieties of melody that have ever greeted the ears of the world, so these four fundamental plans of animalic structure have furnished the endless variations and combinations which daily greet our senses with never-ceasing novelty and delight. As Agassiz has aptly and beautifully illustrated the idea, one of these fundamental plans is like the fundamental harmony upon which an endless set of variations may be played. Vary it to what extent you will, the characteristics of the theme continually recur. What are the zoological characters of these four fundamental forms may be learned from any elementary work on the science. It is the magnificent generalization—for which we are indebted to the genius of George Cuvier—that I wish to impress. Suffice it to say that all animals are either *vertebrated*—possessed of a backbone; *articulated*—with an external horny crust, composed of rings, like insects, lobsters, and worms; *molluscous*—with soft bodies like slugs, very often covered by a shell, like snails and oysters; or *radiated*—with bodies composed of parts somewhat symmetrically arranged on all sides with reference to the centre, like the starfish and corals. I have named the most striking character which distinguishes each of these great branches of the animal kingdom. All the other parts conform to these; indeed, the basis of each peculiar plan is laid in the nervous system, at a very early period of embryonic development; and the hard parts—the bones and external crust—are moulded to this, so that, though the real basis of these distinctions is hidden from view, the external form and proportions become always an infallible exponent of the fundamental plan.

Three of these fundamental plans are called into requisition in the constitution of the very first population of our globe, omitting any consideration of the little-known existences of the Eözoic Time. The coral was a radiate; the *Lingula* was a mollusc; the trilobite was an articulate. The fourth plan was drawn upon before the close of the first great period of animal history, and was realized in the form of a fish.

In the very first chapter of the book of Nature, then, we read the announcement of a programme which is still in process of execution. The type of the primeval coral has sprouted into the sea-anemone, the sea-nettle, and the starfish. The type of the *Lingula* has been degraded into the Bryozoan and nummulite, and expanded into the clam, the snail, and the cuttle-fish. The type of the trilobite has varied into the worm below and the insect above; while the vertebrate type, beginning with the fish, has developed into the reptile, the bird, the quadruped, and man.

Nor does method end here; nor the method which had its first announcement on the morning of animal existence. I have already alluded to the varied conditions under which animal life presents itself—the various ends with reference to which animals have been modified—some to swim, some to fly, some to climb, some to burrow; some for exalted powers and active habits, others for a degraded and sluggish existence. Each fundamental type has been moulded, and warped, and adapted to these varied ends and conditions of being. At the same time, the grand characteristics of the type have been conserved even in the extremest modifications. The modifications of the fundamental plan to adapt it to these various ends are class-characters; and we thus find that Nature has herself grouped the members of each branch into classes. This method is as old as the animal creation. Not only did each creature which played its part in the primordial fauna conform to one of the four fundamental types of structure, but it also conformed to the characteristics of one of the preconceived class-modifications of that type.

Lastly, each class-group is composed of different grades of animals, constituting so many different orders within the limits of the class. This gradation of ordinal types was also recognized in the organization of the earliest animals.

Thus the whole plan of creation was mapped out to the mind of the Creator in the beginning. We shall see, as we proceed with our sketches of the history of creation, that every step in the evolution of continents, and the establishment of a home for the coming man, was a movement in a definite direction, effected by forces chosen from the first, and shaped always with reference to exigencies which were to arise in the far-distant future. We shall see how the simple animal forms of the primeval ocean embodied in themselves germs which were capable of unfolding into the richest variety of adaptations and the most exalted capabilities. There can be no nobler, no more instructive and inspiring employment, than to stand where we do, at the end of this long history, and, looking back upon it, catch its method, and reproduce in our own minds the sublime conceptions of the Architect of the World.

CHAPTER X.

THE GARDEN OF STONE LILIES.

WE have wandered down through the fiery mazes of the præsedimentary ages of the world, and have seen the granite, the quartz, feldspar, and mica, the hornblende, and other first-born products of primeval refrigeration organizing themselves in obedience to the molecular forces of Nature; we have witnessed the floods descending, and cubic miles of sediments settling in the bed of the Eözoic sea; we have gazed upon the first flickerings of animated existence, and have noted the fact that while Nature established the procession of organic being with the four sub-kingdoms of animals nearly abreast of each other, the van of each was led by some of the weakest and most abnormal forms which have ever appeared within the circle of their respective types.

The conditions of existence during the St. John's and Potsdam periods must have been somewhat uniform under all meridians. No continents existed to divert the tidal current into cooler or warmer latitudes, or unequalize the temperature of the atmosphere by their superior power of absorbing and radiating heat. The leading types of existence were trilobites—exhibiting a close relationship with each other on whichever side of the world we exhume their mummied forms—and some inferior brachiopods, which are almost identical in species at St. Petersburg, and at Keeseville, New York. We have seen that the central portions of the American continent constituted at this time a vast basin of shallow water, the rim of which extended all around the frontier of the Middle and Northern States. In this magnificent lagoon the Iron Mountain of Missouri loomed up, as it now does—an island of metal—the apex of an iron cone, whose base rests broadly and deeply on the molten ocean which floats kingdoms and continents from the past eternity to the future. Around its sloping flanks the sediments of the Potsdam period accumulated in horizontal layers, which to-day may be witnessed abutting against the dark sides of the emerging cone of metal. A few other isolated points had thus early been born from the abyss.

In such a sea—a shoreless sea—lived, and lived in happiness, those problematical forms called trilobites, whose remains have been opened from the solid rocks of Wisconsin, Vermont, Canada, and hundreds of other localities. Rather, on such a submarine *platform* they sported their day, for on all sides—certainly toward the east, south, and west—the waters deepened, as now, to an almost unfathomable depth, to whose dark recesses life never gropes its way (Fig. 38).

In the progress of the earth's preparation this act of the drama closed, and the curtain fell upon the scene. "The curtain rose, and the scene was changed." The beings which teemed in the waters of the preceding epoch were buried in the ruins of a convulsion which marked the advent of a new æon. Not an individual of any of the former species outrode the storm. But the sea is now quiet again —more quiet than before. The waters are clearer. The floor of the ocean has settled a few hundred feet deeper, and the conditions of our planet are changed. Lo! now the clearer and quieter waters are teeming with myriads of new existences, some of which reproduce the family features of the beings of the preceding period, while others are forms now first revealed upon our planet. Whence come these new tribes? A convulsion of nature shuts them off from a lineal connection with the generations of the Potsdam period—a convulsion which moulded the basin of Lake Superior (Fig. 30) and notched its southern shore with Kewenaw Point. We descry, moreover, among these new populations forms which could not possibly sustain any genetic relation to their predecessors in the line of being. These all are new creations. There is no avoiding the conclusion. The omnific fiat of the Creator has again gone forth, and swarms of beings innumerable start from the teeming and prolific deep. *Encrinites* now first adorn the flowery chambers of the sea—one of the new ideas just realized from the Creative Mind—flower-like, with slender stem affixed to the submarine soil, a delicate corolla uplifted on its extremity, and petals delicately fringed expanded to the diluted sunlight of the smiling heavens above, struggling down to the coral meadows on which they flourished. And these were *animals.* With all their plant-like form, and grace, and delicacy, and attachment to the soil, these new and wonderful creations had sensibility and will, and enjoyed their allotment at that early age of the world, and at that depth beneath the cheerful sunlight, and the caressing breeze, and the vital air, as the butterfly now, which is borne upon the sunbeam from flower to flower, and sips the sweetest nectar from the fairest creations of the vegetable world. All over the area of the Northern and Western States, and as far south as Alabama and Mis-

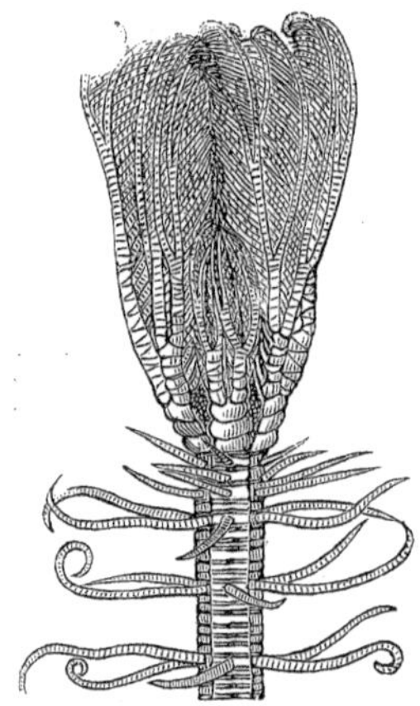

Fig. 40. Pentacrinus caput-Medusæ. A crinoid living in the Caribbean Sea.

sissippi, flourished on the great submarine plateau luxuriant plantations of these little lily-animals. And these were interspersed with other plant-like forms—the coral animals—which reared their marble domes and uplifted their arborescent structures upon the same soil which supported the encrinite and formed the grazing-ground of tribes of molluscous beings.

"Deep in the wave is a coral grove,
Where the purple mullet and goldfish rove;
Where the sea-flower spreads its leaves of blue,
That never are wet with falling dew,
But in bright and changeful beauty shine,
Far down in the green and grassy brine.
The floor is of sand, like the mountain drift,
And the pearl-shells spangle the flinty snow;
From coral rocks the sea-plants lift
Their boughs where the tides and billows flow;
* * * * * * *
While far below in the peaceful sea
The purple mullet and goldfish rove,
Where the waters murmur tranquilly
Through the bending twigs of the coral grove."

Here was beauty, here was sensitive enjoyment, lavished by Nature upon these humble forms at this remote age of the world, and in these "dark, unfathomed caves of ocean," with the same liberal hand which adorns the modern landscape for the admiration of intelligent man. Here again were trilobites—not the same species as had been swept from being by the convulsions which marked the close of the last epoch—but articulated animals, conformed to the same family plan and features as their extinct predecessors, yet as easily distinguished as a wasp from a bumble-bee. And what, still,

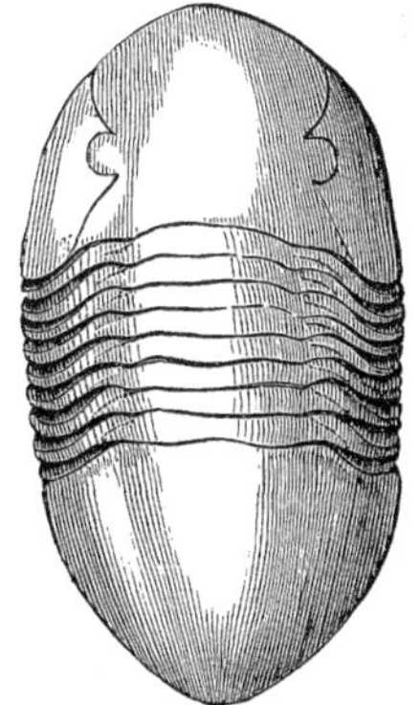
Fig. 41. Asaphus gigas of the Trenton period.

are these new and anomalous forms, which move their sullen and sinister visages among the other tribes with the mien of conscious and insolent superiority? Predaceous creatures, they despoil at a meal the most beautiful bed of encrinites, while the trilobite, alarmed, shoots with a quick stroke of his tail under cover of some coral crag. These are *Orthoceratites.* They were so numerous and powerful, being, withal, the monarchs of the period, that we must pause to look into their family connections.

CHAPTER XI.

THE FAIRY SAILOR AND HIS COUSINS.

WHO has not heard of the argonaut, or paper nautilus? One of the most vivid recollections of our early reading presents us with a little boatman, in his "shelly bark," wafted over the placid surface of a summer sea. With tiny sail upraised, the favoring breeze bears him securely onward; but let the winds escape from their Æolian caves, and the billows wake from their liquid slumbers, and down glides our tiny boatman with his shelly bark, and finds a safe retreat among the marble corridors of the millepores and the madrepores. Montgomery, in his "Pelican Island," has thus embalmed the fable:

Fig. 42. The Paper Nautilus (*Argonauta Argo*).

> "Light as a flake of foam upon the wind,
> Keel upward, from the deep emerged a shell,
> Shaped like the moon ere half her orb is filled.
> Fraught with young life, it righted as it rose,
> And moved at will along the yielding wave.
> The native pilot of this little bark
> Put out a tier of oars on either side,
> Spread to the wafting breeze a twofold sail,
> And mounted up and glided down the billow
> In happy freedom, pleased to feel the air,
> And wonder in the luxury of light."

It seems a pity to spoil so pretty a fable, and one, too, that has lived since the days of Aristotle. But the fable of the argonaut has been spoiled by the industry of a lady.

Madame Jeannette Power, a French lady residing in Sicily, has transmitted to the learned societies of Europe accounts of observations made by herself upon the argonaut of the Mediterranean, which prove that the "native pilot" is the rightful and original owner of the "little bark," while the latter, instead of being devoted to the purposes of fairy navigation, is but a coat of mail for protection against ugly foes, and the "two-fold sail" is the "mantle" extended over the animal's back, a secretion from which forms and enlarges the shell with the growth of the animal. The propulsive power of the animal, instead of Æolian breezes, is a jet of water squirted from a tube or "funnel," which, like a rocket-power, drives the argonaut backward; and its "tier of oars" is used with the animal inverted, crawling, like a snail, with his house upon his back.

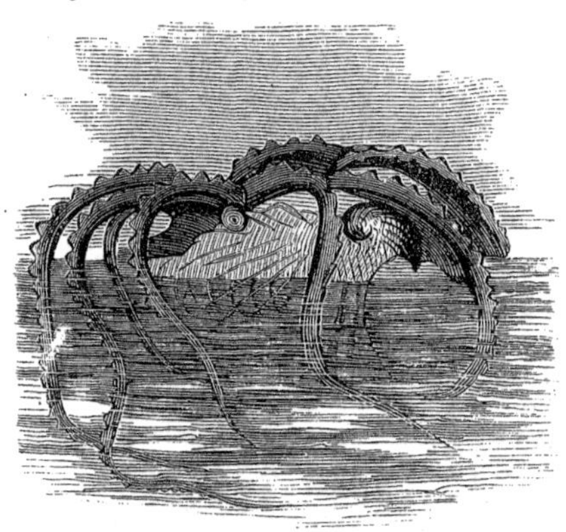

Fig. 43. The Paper Nautilus (*Argonauta Argo*), with the arms of the animal extended.

Something still more familiar to every reader is the "cuttle-fish bone," which the apothecary sells for canaries. This substance is not a "bone," and does not come from a "fish," but is a rudimentary shell formed beneath the skin which covers the back of a molluscous animal. The calamaries are similar to the cuttle-fishes, but their shell is horny instead of stony. The poulp, or cuttle-fish of the southern coast of Europe, has been longest known. It was called "polypus" by Homer and Aristotle, because it has many feet or arms. The aspect of all these animals is strange and uncouth (Fig. 44). Their staring eyes, their

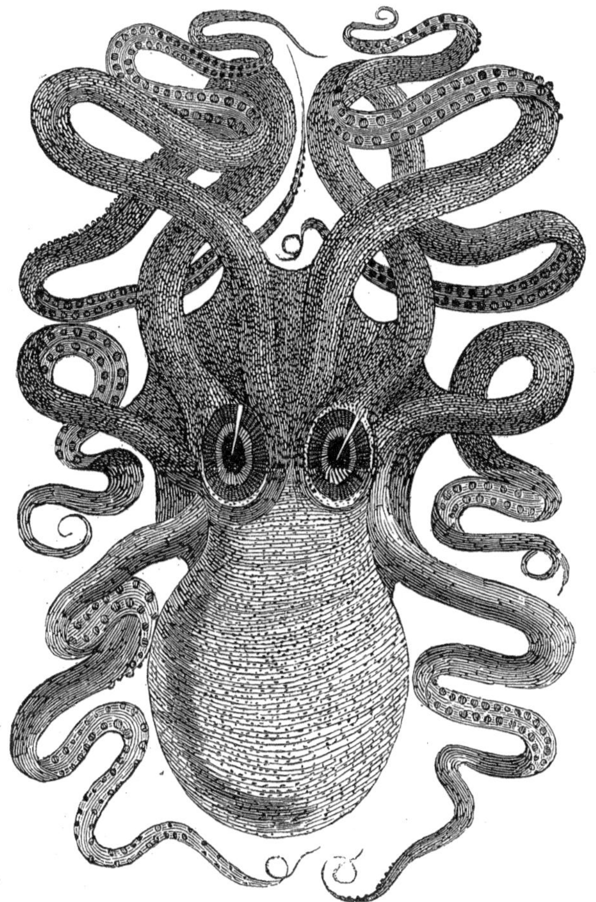

Fig. 44. The Eight-armed Cuttle-fish (front view).

long and flexible arms, and their formidable pair of sharp and horny mandibles, combine to render them unpleasant

Fig. 45. Fac-simile of the Commemorative Painting in the Church of St. Malne, France.

neighbors. Surrounding the mouth is a circle of eight strong arms many times the length of the body, while staring out from either side of the head, between the bases of the arms, is a pair of large glassy eyes, which send a shudder over the beholder. At the bottom of the sea the poulp turns its eight arms downward, and walks like a huge submarine spider, thrusting its arms into the crevices of the rocks, and extracting thence the luckless crab that had thought itself secure in its narrow retreat from the

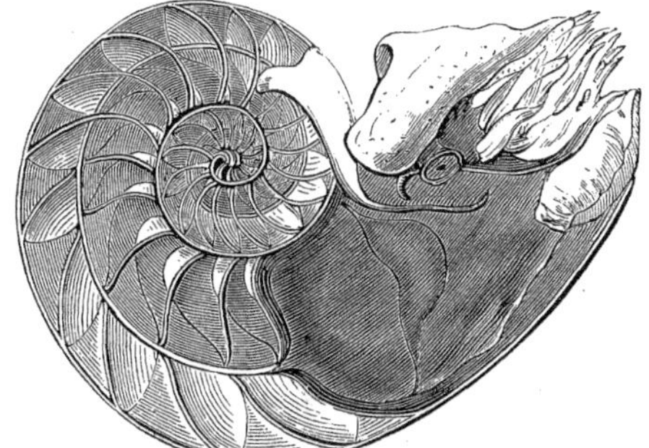

Fig. 46. Section through the shell and animal of the Pearly Nautilus (*Nautilus pompilius*).

attacks of so bulky a foe. This is the "Devil Fish" so graphically but so unscientifically described by Victor Hugo. Each of the arms is covered with what are called suckers, designed for producing adhesion to the object grasped. Each sucker consists of a little, elevated, circular horny ridge, forming a little cup, closed at the bottom by a flexible membrane which is attached to the arm by a stem. The consequence is, that when the arm is pressed upon an object, the bottom of each cup, like a piston, is pressed inward by the action of the stem or piston-rod, which is moved by the pressure of the arm. The effort to escape from the grasp of this arm withdraws the piston back to the bottom of the cup, thus producing a vacuum within, and causing a suction which effectually retains the object. Could any piece of mechanism be more admirable?

The poulp, also called octopus (eight-footed), sometimes attains a formidable size, and sailors relate terrible stories of those found in the African seas. According to Denys de Montfort, Dens, a navigator, avowed that while three of his men were engaged in scraping the side of the ship, one of these monsters reached up from the water its long and flexible arms, and drew two of the men into the sea. One was never rescued, and the other, after his escape, became delirious and died. This was probably a "sailor's yarn," since the Frenchman who narrated it afterward represented a "Kraken octopod" in the act of scuttling a three-master (Fig. 45), and told M. Defrance that, if this were "swallowed," he would, in his next edition, represent the monster embracing the Straits of Gibraltar, or capsizing a whole squadron of ships. Little reliance as can be placed in the marvelous stories of "those who go down to the sea in ships," it is well authenticated that some of these octopods attain fearful dimensions, being the largest invertebrates known. Milne-Edwards, an eminent Parisian naturalist, has expressed the conviction that the unexplored depths of the ocean conceal the forms of octopods that far surpass in magnitude any of the species known to science.

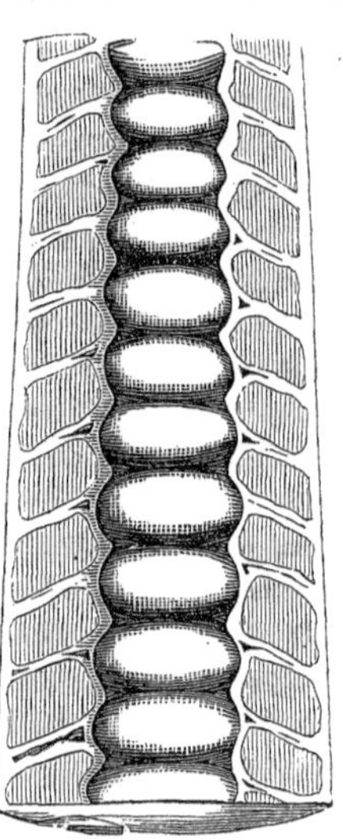

Fig. 47. Fragment of a straight-chambered Shell (*Ormoceras tenuifilum*), showing a large annulated central siphon.

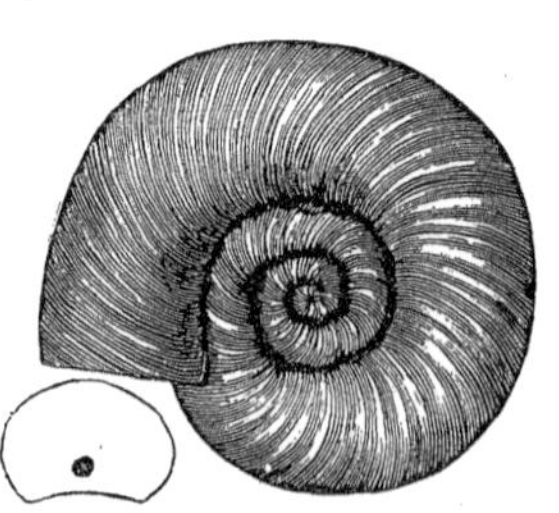

Fig. 4. Trocholites ammonius. A coiled-chambered shell of the Trentón period.

The common cuttle-fish of our own coast is a much more harmless animal, attaining a length of only ten or twelve inches. The calamary of New York Harbor has ten arms, two of which are much longer than the others.

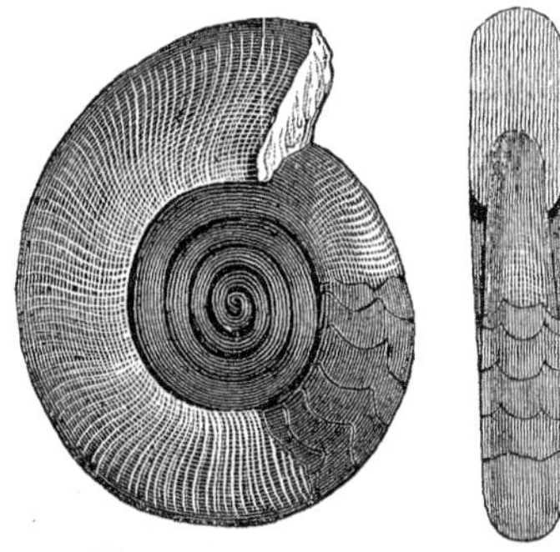

Fig. 49. Clymenia Sedgwickii.

The reader is probably familiar with the sepia used in tinting with water-colors. This is the ink of the cuttle-fish and its allies. It is preserved by the animal in a little bag, from which it is ejected on the approach of danger, thus producing a cloud, under cover of which the animal escapes. Here is the prototype of the fog which sophistry raises, and under cover of which it retreats, when finding itself in unequal conflict with truth. India ink, it has been stated, is manufactured by the Chinese from the same substance, though it is probable they employ only lampblack and glue, or vegetable gum. The ink-bags of some ancient cuttle-fishes have been found in a fossil state. Dr. Buckland had drawings of extinct species executed in their own ink.

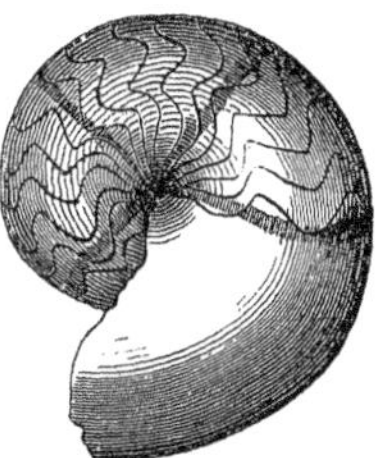

Fig. 50. Goniatites Allei (from the Marshall Group, Michigan).

These all are cephalopods, the first class among molluscs, the aristocracy of shellfish, often exercising dominion over beings with higher intelligence, but a weaker arm, just as brawny force has always done. But the forms described belong to the highest of the two orders of the class.

None, save the "paper nautilus," have had external shells. The animals of the lower order are incased in shells which are long, tapering tubes, divided at regular intervals by transverse partitions. The paper nautilus and his allies have all lived in a later age of the world than that of which I have been speaking. The "*pearly* nautilus" is the only living representative of the lower order—an order which swarmed in the seas of the Paleozoic and Mesozoic Times. The pearly nautilus is closely coiled (Fig. 46); its shell is divided at frequent intervals by smooth partitions concave anteriorly, the animal occupying only the space in front of the last one. A shelly tube runs through the middle of all these chambers to the farther extremity of the shell. Through this a ligament passes from the body of the animal, and anchors it securely in the last chamber. This tube is called the *siphon*. Such is the structure of the pearly nautilus, which may be seen in myriads, on a calm day, floating on the surface of the waters of the South Pacific.

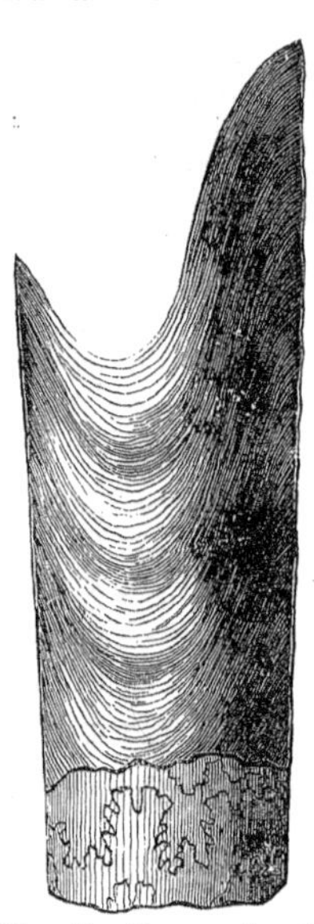

Fig. 51. Fragment of Straight-chambered Shell (*Baculites ovatus*) of Mesozoic Time, belonging to the Ammonite family.

The reader will certainly thank me for introducing here a beautiful poem on "the Chambered Nautilus," though the author has committed the error of supposing it was this species of nautilus to which the Aristotelian fable of the fairy sailor applied.

> "This is the ship of pearl, which, poets feign,
> Sails the unshadowed main—
> The venturous bark that flings

On the sweet summer wind its purple wings,
In gulfs enchanted, where the siren sings,
And coral reefs lie bare,
Where the cold sea-maids rise to sun their streaming hair.

"Its webs of living gauze no more unfurl;
Wrecked is the ship of pearl!
And every chambered cell,
Where its dim dreaming life was wont to dwell,
As the frail tenant shaped his growing shell,
Before thee lies revealed—
Its irised ceiling rent, its sunless crypt unsealed!

"Year after year beheld the silent toil
That spread his lustrous coil;
Still, as the spiral grew,
He left the past year's dwelling for the new,
Stole, with soft step, its shining archway through,
Built up its idle door,
Stretched in his last found home, and knew the old no more.

"Thanks for the heavenly message brought by thee,
Child of the wandering sea,
Cast from her lap forlorn!
From thy dead lips a clearer note is born
Than ever Triton blew from wreathéd horn!
While on my ear it rings,
Through the deep caves of thought I hear a voice that sings:

"Build thee more stately mansions, oh, my soul,
As the swift seasons roll!
Leave thy low-vaulted past!
Let each new temple, nobler than the last,
Shut thee from heaven with a dome more vast,
Till thou at length art free,
Leaving thine outgrown shell by life's unresting sea!"

The "pearly nautilus" exemplifies the structure of a "chambered shell." Such shells in their endless variations played a most conspicuous part in the history of ancient life, though one genus alone survives to recite the glory and illustrate the economy of his cephalopodous ancestors. The variable elements in the shell are the form of the septum, the position of the siphon, and the plan of enrollment. The septum may be plain, or angulated, or lobed, or foliated around its outer margin. The siphon may be external, or internal, or central. The enrollment may be close, loose, half-coiled, arcuate, or straight. Of how many combinations, three in a set, do these characters admit! And yet almost every possible combination has been realized in the history of the world. In the earliest periods were the species with simple septa and straight shells (*orthoceratites*, Fig. 47); next came those with simple septa and coiled shells (*Nautili*, Fig. 48); then those with angulated septa and coiled shells (*Clymenia*, Fig. 49); then those with lobed septa and coiled shells (*Goniatites*, Fig. 50); lastly appeared those with foliated or very complicated septa, with their straight (*Baculites*, Fig. 51), arcuate (*Hamites*), closely (*Ammonites*, Fig. 52), and variously coiled forms. So we see that in the various ages of the world, some type of "chambered shells" has constituted a leading character-

Fig. 52. Ammonites canaliculatus. A chambered shell of the Mesozoic Ages.

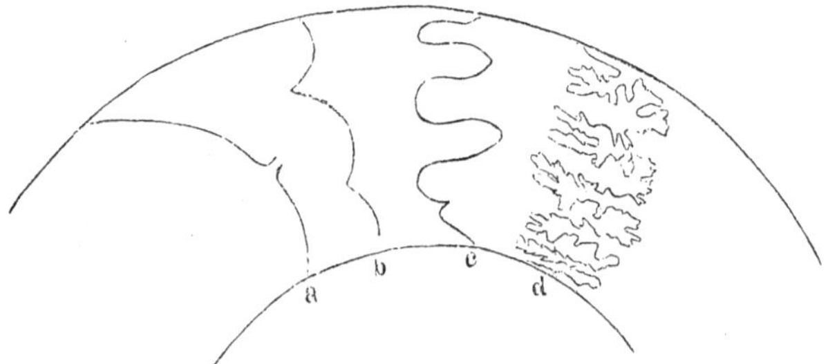

Fig. 53. Plans of Septa among different families of Chambered Shells.
a. Septum in Nautilus family. b. Septum in Clymenia family. c. Septum in Goniatite family (*Goniatites Marshallensis*). d. Septum in Ammonite family.

istic of the marine fauna. One thing which is very remarkable is the fact that the existing pearly nautilus is closely related to the most ancient forms—a specimen creature of primeval times—the key to the inscriptions on the preadamite rocks. The orthoceratites were nautili with straight shells. They were the "carnivora" of the sea. They often attained to formidable dimensions. I have found remains of individuals on St. Joseph's Island, in Lake Huron, which were twelve feet in length. A reliable gentleman of Utica, New York, informed me that he had traced one in the "Black River Limestone" to the distance of thirty-two feet! Imagine a hollow cone of limestone, of the dimensions of a "saw-log," animated, with a "Kraken octopod" ensconced in the open end, staring with glassy, sinister eyes to the right and left, and numerous slimy, muscular, insinuating arms feeling in every direction for their prey.

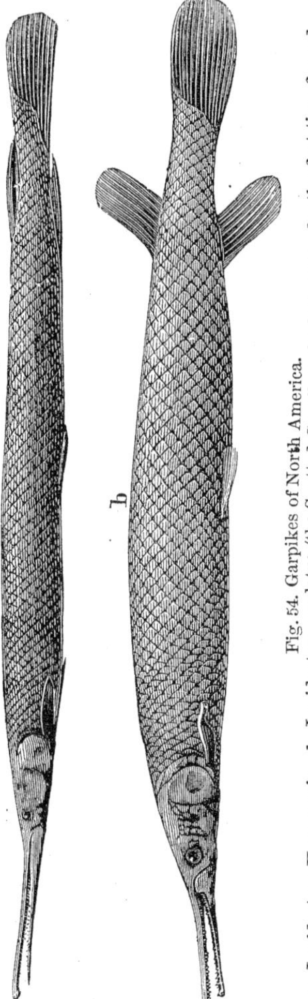

Fig. 54. Garpikes of North America.
a. Lepidosteus Huronensis. b. Lepidosteus oculatus (the Spotted Garpike), the latter now for the first time figured.

Is not this an enemy from which the lesser tenants of the deep would flee without pausing to raise the question of supremacy? These monsters maintained the ascendency till the introduction of fishes, toward the close of the Upper Silurian, or later. Their decline dates from this epoch; and when the voracious fishes of the Old Red Sandstone and the Carboniferous Limestone came upon the stage of being, the orthoceratites dwindled away. Their last representatives barely saw the rising of Mesozoic Time. Not a trace of a straight plain-chambered shell has been found in any of the rocks above the Trias. They fulfilled their end in creation and retired. Other carnivorous animals of a higher order were better adapted to the advancing state of the earth's preparation. Garpikes appeared. A new dynasty arose, to be in turn overthrown by the dynasty of the Mesozoic reptiles.

CHAPTER XII.

ONWARD THROUGH THE AGES.

THE evening shades of one of eternity's æons are gathering around us. The darkness upon which we are entering is the gloom of a tempest and the night of death to the teeming populations of the globe. A throe of Nature heaves still higher the germinal ridges of the continent, robs the ocean of another strip of his domain, and seals up the record of the life of the Lower Silurian.

The elevation which marked the close of this great interval of terrestrial history brought to light the basin of Lake Superior, Northern Wisconsin, and Minnesota, the northern and eastern portions of New York, and considerable portions of New England. The line of sea-coast passed westward through Central New York, along the bed of the future Lake Ontario, thence northwestward to Georgian Bay, following the trend of the future Lake Huron, sweeping round by the Sault Ste. Marie, and arching downward again through Wisconsin along a line a few miles west of the present Lake Michigan. Thence it swept westward and northwestward in the direction of Lake Winnipeg and the Arctic Sea. All to the south of this line was yet the empire of the Atlantic. On those vast submarine plains the Pacific joined hands with the Atlantic, and the two sang dirges over the land that was to become the scene of fraternal conflict.

It might weary the casual reader of geological history to recite the details of the periods which follow. What has been narrated of the birth and death of populations

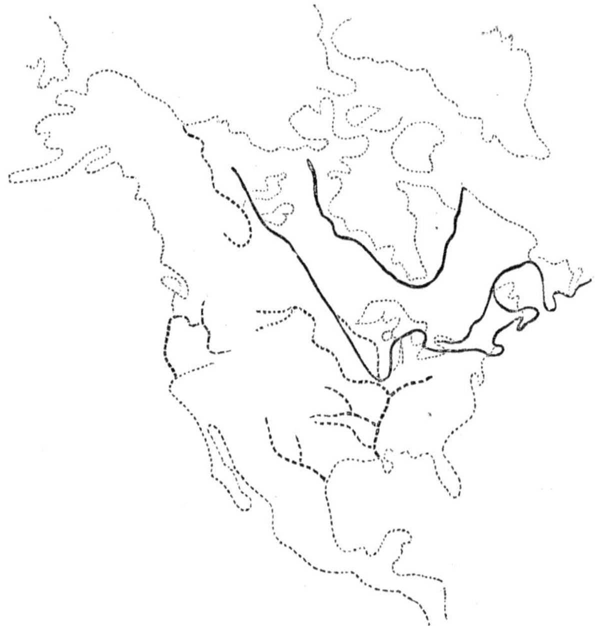

Fig. 55. What the North American Continent had become at the end of the Silurian Age. (The modern continent is indicated by dotted lines; the rivers by broken lines.)

during the Lower Silurian Age will answer for a representation of the nature of the events which followed during the Upper Silurian and Devonian ages. Successive extinctions, wrought by the lapse of time, or by violent geological revolutions, followed by successive creations of higher and higher forms, and the annexation of successive belts to the pre-existing land—these constituted the great secular features of the world's history down to the dawn of the period when air-breathing animals were to have birth (Fig. 56).

The first period of the Upper Silurian was that during which the Niagara limestone was accumulated—a forma-

Fig. 56. A remarkable Silurian Sea-weed (*Arthrophycus Harlani*). From the Medina Sandstone of the Niagara Group.

tion through which, with others, the Niagara River has cut its way. In another connection I shall have something farther to say in reference to this stupendous piece of Nature's engineering. From the falls of Niagara the outcropping belt of this limestone runs in lines parallel with those just traced. It forms the promontory of Cabot's Head, and the peninsula separating Georgian Bay from Lake Huron. At this point the formation has succumbed to the attacks of the waves, and disappears in its northwestward trend beneath the water of the lake. Cropping out again, it forms the remarkable chain of the Manitoulin Islands, in the northern part of Lake Huron, including Drummond's Island. Beyond St. Mary's River it forms a "point" and a peninsula, the counterparts of Cabot's Head and the peninsula to the south of it. Running westward, and then southwestward, it establishes a continuous barrier to Lake Michigan along the northern and western borders, constituting the rocky ridge which isolates Green Bay and Bay de Noquet from the greater lake. It follows the shore of Lake Michigan to Chicago, and even to Joliet, when it bends westward and northwestward, and loses itself beneath the accumulations of a later period. The quarries at Lockport, New York, and many others in that vicinity, are located in this important limestone. In the same formation are those at Milwaukee, Waukesha, Chicago, Lamont, and Joliet. The so-called "Athens Marble," so extensively employed in Chicago, is quarried from this formation. It much resembles the famous "Kentucky Marble," from which the beautiful monument and statue to Henry Clay, at Lexington, is built—though the latter comes from the Trenton group, in the Lower Silurian.

The second period was that of the Salina group, which has become famous for the production of salt and gypsum, in the vicinity of Syracuse, New York. Its outcropping belt runs in a line parallel with that of the Niagara limestone throughout its whole course, as far as Milwaukee. I shall hereafter offer some explanation of the circumstances under which salt and gypsum have accumulated to such an enormous extent in certain formations.

The third period was that of the Lower Helderberg group, which is not found to be generally spread out over the country like the other two. In New York it is especially developed in the Helderberg Mountains, where Professor Hall has obtained a rich harvest of organic remains. It was here that he found the type of that magnificent crinoid, which he so beautifully named *Mariacrinus*, in commemoration of the assistance and sympathy of his accomplished wife in his life-long scientific labors. It thins out and disappears in Western New York. This group is known again in Southern Illinois, where it has been brought to light by the indefatigable and well-directed labors of Professor Worthen; and in Missouri, where it has been illustrated by Swallow and Shumard; and, finally, in Maine, where it has been studied by Professor C. H. Hitchcock, and Mr. Billings, of Montreal.

Thus closed the Silurian Age. At the east the rocks of this age are marked off from the great mass of overlying Devonian strata by the interposition of a conglomerate—the "Oriskany Sandstone," which signalizes the confusion attendant upon the change of scene. At the West, however, this formation is generally wanting; and we find the limestones of the Corniferous group resting upon those of the Niagara group, except where the Salina rocks intervene. The Corniferous is a most important limestone mass throughout the West. It merges generally into the calcareous portion of the overlying "Hamilton" strata, and forms a landmark in the topography of the country no less than in the series of rocks. In this limestone, quarries are worked from Western New York, in the latitude of Buffalo, through the contiguous peninsula of Canada to Sandusky and Columbus, Ohio, Monroe and Mackinac, Michigan, and multitudes of points in Indiana, Illinois, and Iowa. These limestones, like all others, were accumulated in the bottom of deep and quiet seas. Each successive floor has been the home of moving myriads of sensitive forms. Every layer of rocks has been the cemetery of many generations. Life teemed especially in calcareous and placid waters. Such were those of the Corniferous period; and these limestones are stocked with the relics of ancient dynasties—great and small, powerful and weak, in one wide burial confusedly blent. Nor yet had nature dispensed with the pattern of the trilobites. Encrinites were still in vogue, and orthoceratites, and all the various phases of univalve and bivalve creation. And here—here first dawned upon our planet an animal with a backbone—a mere fish, but yet the basis on which artist Nature has moulded successive models till the form of man shone forth, and the Omniscient was satisfied to stay his hand. But man was not yet. Ichthyic life seems to have dawned upon our earth in remarkable profuseness. The bones, and plates, and jaws, and teeth of fishes large and small have been cleft from the Corniferous limestone in Canada, Michigan, and Ohio. Our first authentic information of these earliest vertebrates came from Dr. J. S. Newberry—equally distinguished in the service of science and his country—and who has very recently worked up a wonderful collection of Devonian fishes, created mainly by the intelligent industry of a German Methodist minister, Rev. Herman Herzer, while discharging the duties of his ministry at Delaware, Ohio. These ancient fishes were only the *avant-couriers* of the shoals of sharks, and sturgeons, and garpikes which made a Golgotha of the Old Red Sandstone.

The closing convulsions of this epoch upheaved still higher the growing continent, and depopulated the coral cities of the sea that had just been astir with being. A pause, and another epoch—the Hamilton epoch—followed, a period characterized by its abundance of argillaceous sediments, and by two masses of black bituminous shale—the "Marcellus" at the bottom, and the "Genesee" at the top, with the more calcareous strata between. The absence of the "Marcellus" at the West has dropped the limestones of this group upon the top of the Corniferous limestone, and formed the appearance of but a single mass. This is clearly seen in the extensive quarries upon the islands in the western part of Lake Erie. Indeed, the absence of the "Oriskany" at the West has brought the calcareous portions of four groups of rocks into immediate juxtaposition. These are the Niagara, the Salina, the Corniferous, and the Hamilton. Before these groups were correctly discriminated, the entire mass was known in the West as the "Cliff Limestone." No epoch of the world's history ever witnessed a greater profusion of life than the Hamilton. The germs of being were thickly strewn over every part of the ocean's floor. Chambered shells were on the wane; but Brachiopods and new forms of corals sprang forth in exuberant growth, and we pick their fossil forms to-day, like nuts, from the dried ocean mud.

Another æon passes; the empire of the sea crumbles before the conquest of the land, and we add next the belt of the "Chemung group" to the growing margins of the land. Toward the west the bottom of the sea experiences at this time but little change of level, and the Chemung sediments abide another epoch to receive upon their backs the sands and mud of the "Waverly group;" eastward, however, new land is made by an extensive uplift of the sea bottom. Thus the Empire State is almost completed; Wisconsin has taken her place; the centre of Michigan is occupied by an inland sea. The great ocean washes the southeastern shores of Ohio, and wild waves career over the future plantations of the prairie farmer in Illinois. Some parts of Eastern Iowa, and Missouri, and Arkansas, and Northern Texas begin to emerge, but the boundless waste of Pacific waters is still at work upon the materials of Kansas, and Nebraska, and the regions beyond.

Among the accumulated treasures of this epoch, behold the first vestiges of an arborescent vegetation! All before this had been fucoidal in its characters. Here we find, imbedded in the friable sandstone, some stems of trees—pieces of drift-wood floated from some neighboring shore, and, like the dove of Noah, bearing us tales of the vegetation upon the land. The sandstones of Southern New York inclose such records of the vegetal life of the Chemung. Corresponding sandstones in the distant peninsula of Gaspé, Canada East, have been constrained to yield similar testimony from their locked and ancient archives—thanks to the diligence and learning of Dr. J. W. Dawson, of M'Gill College, Montreal. How sparse and desolate must have been those forests! No voice of animated nature was yet heard among those scattered pigmy trees. They are arborescent ferns and lycopodiums—a new idea incorporated in vegetal existence—but how prophetic of that which is to come! Nature always issues her bulletins. We stand now in an age of the world which antedates the advent of all our familiar forms, and read the announcement of the coming riches of the Carboniferous era. A stranded log of drift-wood becomes eloquent in the utterance of prophetic truth.

Another age passed which the scientific world hesitates to attach to the future or the past. Is it Devonian or Carboniferous? Throughout the West the sediments of this age gave rise to a noticeable formation which has been styled the Marshall group, because the characteristic rocks and fossils of the period may be studied at Marshall, in Michigan. This is the rock so extensively worked in the vicinity of Cleveland, and at Waverly, Ohio. It furnishes the excellent grindstones of Berea, and those known as Huron grindstones in Michigan. It is the greenish or reddish-yellow sandstone occurring in Southern Michigan, and trending northward into the bight of the coast which separates Saginaw Bay from Lake Huron. It underlies the limestone bluff at Burlington, in Iowa, and makes itself known at numerous localities throughout the northwestern states. In New York, it is perhaps the formation corresponding to this which caps the Catskill Mountains, and has hence been styled the Catskill group. It covers a large area in Northeastern Pennsylvania. In this formation, throughout its wide extent, are found the scales and teeth of fishes, which recall the relics studied by Hugh Miller in the quarries of Cromarty, and hence we have been inclined

Fig. 57. Ideal Landscape of the Devonian Period.

to believe that it belongs to the same age of the world. But the "Old Red Sandstone" of Miller is generally partitioned off with Devonian strata, while the western beds of the American formation abound in relics which recall the life and times of Carboniferous populations. Indeed, though some excellent authorities persist in pronouncing the Marshall and Waverly rocks as belonging to the Devonian age, there is not a Western geologist who does not believe them Carboniferous. I have myself had the good fortune to study the fossil remains of this age, gathered from all the Western States by my own hands, and to compare them with fossils gathered from the Carboniferous rocks of Europe, and also with fossils from the Catskill sandstone of New York and Pennsylvania, and I have but little hesitation in asserting that the rocks called Marshall and Catskill were both deposited during the period of the "Mountain Limestone" of Europe, which lies at or near the base of the great Carboniferous system.

If, then, the Catskill sandstone be the base of the Carboniferous system in America, the Old Red Sandstone of Scotland, which has been identified in age, must be, contrary to the prevailing opinion, the base of the Carboniferous system in that country. In North America, the sediments of this period were derived from the wear of ocean shores lying toward the northeast of the United States. The coarser materials were deposited near their source, while the finer were distributed over the centre of the continent. Thus the formation, which is a conglomerate or coarse sandstone in New York and on the shore of Lake Huron, is a fine sandstone in Southern Michigan, in Ohio, and Iowa, and an arenaceous or argillaceous limestone in Southern Indiana, Illinois, and Missouri. In the Old World during the same period, the coarse sediments gave rise to the sandstone of Scotland, while in Yorkshire, Belgium, and other southern regions more remote from the northern source of the sediments the rocks of the age are represented by the "Mountain Limestone." [See Appendix, Note IV.]

I am here led to direct the reader's attention to an important law which has governed the distribution of sediments in all the periods of American paleozoic history. The continent, it will be remembered, was always toward the north. Soundings in the North Atlantic indicate that the actual foundations of the continent extend northeastward beneath the water far beyond the limits of the existing land. Far back in the antiquity of our continent the Labrador branch possessed an extent which no longer appears. It projected itself in that direction almost to mid-ocean. It has been eaten up by the waves of the Atlantic. The bones of the continent lie scattered along from the "Grand Banks" to Maine. Newfoundland, Cape Breton, Nova Scotia, and the numberless islands and peninsulas of the northeast coast, are the remnants of the meal which old Ocean has made of the right wing of America. Out of the wasted continent of paleozoic times the agencies of Nature have built up the substructures of the Northern United States. All the strata to which I have referred were formed of the *ruins* of rocks that had long before been dry land. Thus the materials came from the northeast. And thus it happens that every formation is coarser in that direction, and finer toward the centre of the continent. Thus even the age which witnessed the accumulation of pebbles or sand at the East, witnessed the deposition of a fine calcareous mud in the deeper, quieter waters which rested over the Mississippi Valley (compare Fig. 15).

Another thought introduces itself into the company of this one. It is the law of the secular recurrence of identical lithological conditions. This law attracted my attention many years ago, but I believe no one has distinctly enunciated it except that admirable geologist, Dr. Dawson, of Montreal. Geological time has been marked off into Ages, Periods, and Epochs by physical revolutions. These were universal for the Ages, but more local for the subordinate divisions of time. The commencement of every interval of time was characterized, to some extent, by disruptions, upheavals, violence, emission of heat and vapors from beneath the crust, violent dashing of waters against coast-barriers, destructive ocean tides and streams, and the more or less complete extinction of living beings. Simultaneously, therefore, with the disappearance of a fauna from the earth, the ocean's bottom was overstrewn with the coarse *débris* of a geological revolution. As the shaken crust subsided to a more quiet position, only the finer sediments were transported to great distances from the shores. Lastly, when peace and stability were again restored, the vast expanse of the ocean, as it floated over the area of North America, was a calm and clear lagoon, in which lived and labored those lime-loving animals which incase themselves in shells, found coral structures, and eliminate from the water the materials of limestone strata. There is, consequently, for each period of the world's history a definite succession of strata as to kind. These may be designated Coarse-fragmental, Fine-fragmental, and Calcareous. The Coarse-fragmental we style conglomerates, and their position is at the bottom of a group of strata. The Fine-fragmental vary from sandstones to shales, and they rest upon the conglomerates. The Calcareous constitute the limestones which answer to the culmination of a geological interval, and rest near the top of the group. The life of each interval attained its full expansion during the Calcareous epoch. Toward the close of this epoch the waters of the sea began again to be turbid, from the premonitory jarrings which were soon to be followed by a more or less general disruption. We may generally distinguish, therefore, some calcareous shales constituting the uppermost beds of a group; and, in rare instances, the disturbance proceeded so far before the extinction of the faunas that the uppermost beds have been rendered finely fragmental. To illustrate and confirm these generalizations, I introduce the following table:

	Coarse-fragmental.	Fine-fragmental.	Calcareous.	Calcareo-fragmental.
Lower Silur.	Potsdam sandstone.	Calciferous and Chazy formations.	Trenton Group.	Cincinnati Group.
Upper Silurian.	Oneida conglomerate. Medina sandstone.	Clinton Group. Niagara shale.	Niagara limestone.	Salina Group.
Devonian.	Oriskany sandstone.	Schoharie Grit.	Corniferous limestone.	Hamilton Group, followed by Chemung.
Lower Carb.	Waverly sandstone (Marshall phase).	Waverly sandstone (Chouteau phase).	Mountain limestone.	False coal-measures.
Upper Carb.	Parma conglomerate.	Coal-measures (broken into many short epochs).	Laramie limestone.	Permian Group.

In this exhibit I take no account of the St. John's Group, since we know so little of its lithological characters. It thus appears that the recognized succession of strata in each of the great divisions of Paleozoic time is wonderfully similar in lithological characters. In each great group is a great limestone mass, which stands out conspicuously in the geology of the region underlaid by the group. These limestone masses are prominent landmarks in the progress of the ages. They mark the successive culminations of the geological periods. Each mass outcrops in a protruding belt, sweeping from east to west over a wide extent of country. The oldest is the more northern, and the others follow in regular succession. The Trenton mass sweeps across along the north of Lake Ontario and to Georgian Bay. The Niagara mass lies to the south of Lake Ontario

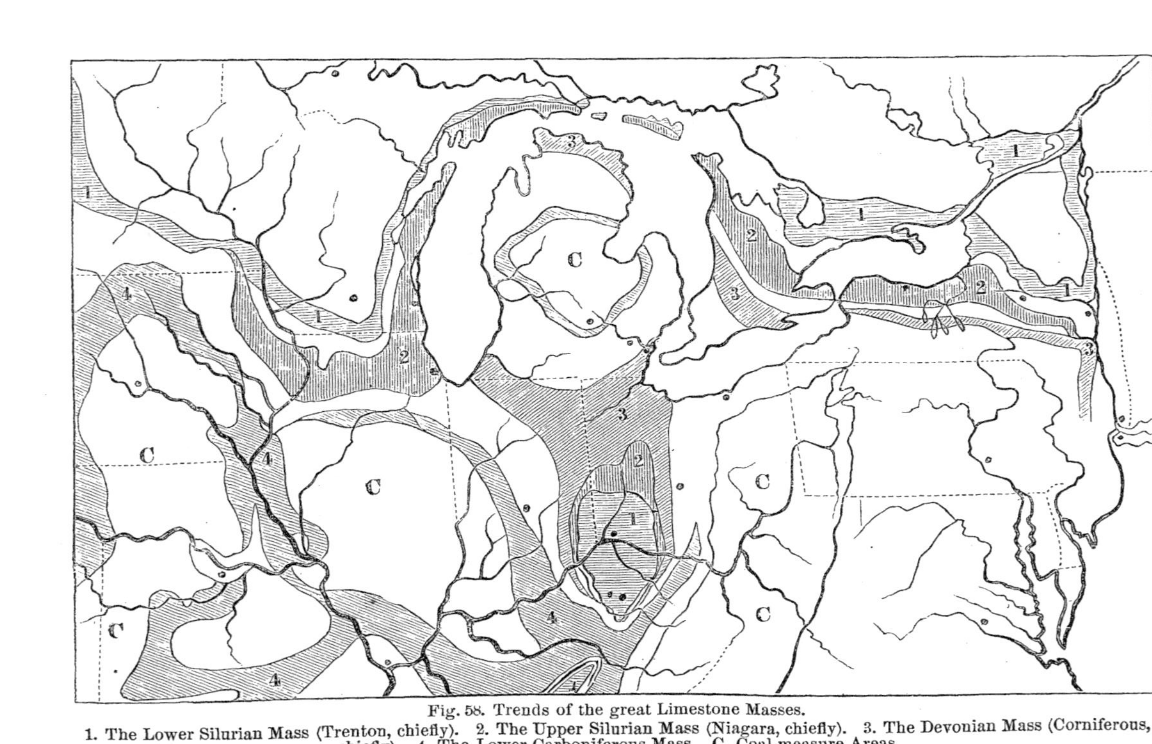

Fig. 58. Trends of the great Limestone Masses.

1. The Lower Silurian Mass (Trenton, chiefly). 2. The Upper Silurian Mass (Niagara, chiefly). 3. The Devonian Mass (Corniferous, chiefly). 4. The Lower Carboniferous Mass. C. Coal-measure Areas.

and the south of Georgian Bay. The Corniferous is north of Lake Erie and beneath Lake Huron. The Mountain limestone is farther toward the centre of the continent, in the Mississippi Valley. The Laramie limestone stretches to the Rocky Mountains. If the reader can fix his imagination on each of these great limestone belts, he has a clew to a mental map of the geology of the country.

In the little map on the preceding page I have endeavored to indicate the locations of the great limestone masses just alluded to (except the Laramie limestone, which is too far west). The *horizontal shading* shows the trend of the Lower Silurian mass, which, in Ohio and farther west, is not discriminated from the Cincinnati Group. Its prolongation into Wisconsin is covered up with surface sands and clays. The *vertical shading* indicates the trend of the Upper Silurian mass, which is also lost in Wisconsin. In Ohio it probably exists in a belt encircling the Lower Silurian area, but it has not yet been completely traced out. The *oblique shading from right to left* denotes the great Devonian mass (corniferous limestone), which has not yet been distinctly traced beyond Lake Michigan. The *oblique shading from left to right* is the Mountain limestone, or Lower Carboniferous mass, which I have proposed to designate the Mississippi Group, because so extensively developed in the valley of the Mississippi River. Now, if the reader desires to know to what particular formation any proposed limestone quarry belongs, this little map will inform him. The letter C indicates the areas which are underlaid by the coal-measures of the country. In the Northern States these are the uppermost strata of solid rock. Hence all other formations *dip* toward the nearest coal-measures, and generally pass under them. In other words, all the strata numbered from 1 to 3 dip toward the areas marked C. It follows, also, that Nos. 2 and

3 dip away from No. 1. No. 4 and C are generally nearly horizontal, except in the vicinity of mountain disturbances. Now, by keeping in mind the dips of the several strata, and tracing each, in conception, underneath those which cover it, the reader will be able to present to his imagination a sort of stereoscopic view of the underground structure of the Northern States.

I have now—leaving out of the account the debatable

Fig. 59. Hugh Miller, the Scotch Geologist.

Marshall—barely enumerated the epochs of two great periods of the world's organic history, the Silurian and Devonian. Who has considered the measureless intervals which have been so glibly hurried over—the rising and setting suns, the passing tempests, the lonely-budding tree, the sands worried to and fro upon the uncertain beach, the lives of myriads of conscious forms in a long succession of populations, the heaving shore, the rise of continents, the burial of beautiful but senseless ruins beneath acres of sediments from which they shall never be exhumed? Let me commend the sublimity of the theme to the reader's attention.

We are now on the threshold of another great period of the world's history. Graceful tree-ferns are waving in the distance, and giant club-mosses are uttering from their fronds a breezy murmur refreshing to the mind wearied with the contemplation of the uncouth and sombre forms which vegetated in the earlier seas. Looking through the vistas of the future, we behold lazy reptiles reposing upon banks protected by the tangled stems of lepidodendra and calamaria, or floating in the tepid bayous of a tropical jungle. The novelty and interest of the prospect invite us onward, but the vastness of the field bids us pause and refresh ourselves before we venture upon our jottings from the scenes of the Carboniferous Period.

CHAPTER XIII.

AN UNDERGROUND EXCURSION.

FOUR hundred feet beneath the foundations of the city, with its piles of brick, and marble, and iron—beneath the roots of the oaken forest and its Dodonean colonnades—beneath the bed of the flowing river and its freight of animated hulls—down four hundred feet beneath the light of the nineteenth century, guided only by the glimmer of the oil lamp suspended from his smutty cap, the miner works the coal which blazes in the cheerful grate, or wakes the slumbering energy which drives the monster steamer on the stormy wave. Let us enter the yawning avenue to this subterranean world. [See Appendix, Note V.]

Armed each with a miner's lamp, and clad in a miner's garb borrowed for the occasion, we step upon a platform, or "cage," six feet square, suspended by iron rods connected with machinery moved by an engine, and, at the word, begin to sink into the gulf of blackness beneath us. This perpendicular hole, perhaps eight feet square, is called the "shaft." By the light of the outer world thrown into the mouth of the chasm, we perceive that the shaft passes at first through a few feet of sand and gravel. Lower down the darkness of the pit enshrouds us, but we learn by the gleam of the lamps that we are passing through fifty feet of coal-black shales, which, like the sandy beds above, are held in their places by a frame of planks. We next find ourselves in the middle of an aperture through a bed of limestone perhaps twenty-five feet thick. The walls are

Fig. 60. Miners going down a Shaft. After an engraving by Bonhommé.

studded with the shells of molluscs which lived and enjoyed existence when this limestone was the ocean's bed, and the light of day shone down upon their quiet abodes as it now shines upon the busy builders of the coral reef. The light of day!—but a day of God's eternity, which dawned upon our planet before Elohim had said, "Let us make man in our image." Rapidly through the belt of limestone our little car descends, and we next find ourselves environed by a wall of sandstone. Here and there are streaks and patches of dark carbonaceous material, and occasionally the eye catches glimpses of woody stems imbedded in the solid rock. But hark! a sound of water rises from the darkness beneath. A subterranean stream has been intercepted, and a little rill is trickling down the massive wall-side. Again in the midst of black, bituminous shales; and now we hang suspended opposite an opening in the stony wall. One hundred feet above our heads the light of heaven is still visible, and three hundred feet below are darkness and emptiness. On the right and the left are entrances to chambers which have been excavated in a seam of coal occurring at this level. But the end of our journey is not here. Continuing to descend, we perceive the bed of coal underlaid by clay, with abundant grass-like shoots and occasional stems of vegetation. In turn we pass shales and sandstones, and then seams of coal, till, at the depth of two hundred feet beneath the surface, we hang before another portal to a long, dark avenue excavated in a deeper-seated bed of coal. In some of the dark and dusty chambers of the labyrinth which opens here the miner's pick is heard resounding, and now and then the muffled report of the miner's blast comes echoing through the vaulted aisles. But this is not the station where we intended to stop. Our car moves on, and we plunge through two hundred feet more of the

Fig. 61. View in an English Coal Mine.

rocky rind of the earth. Above us, the mouth of the shaft seems narrowed by perspective into an insignificant hole; before us opens a dark street, over which, on a tramway, mules are hauling car-loads of coal, which is starting on its journey to the populous city (Fig. 61). Miners, with their picks, are moving to and fro; the sound of hammers is heard; the paraphernalia of busy life are about us, and we seem translated to a nether world. We feel like the hero of the Latin song, who got permission to visit the realm of Pluto, and make the acquaintance of unborn spirits destined to dawn upon the world in the coming Golden Age. Where is the Styx and its sleepy boatman? Where are the shades that expectation thinks to see flitting before us? Let us enter this dingy street, and conjure spirits from their Lethean sleep upon the coaly couches that line the passage-way.

The seam of coal is a broad, horizontal sheet or bed from three to five feet thick. In this are excavated passages about eight feet wide and about five feet high. A main "gangway" may be half a mile or a mile in length. From this, at suitable intervals, lateral passages or "chambers" are quarried out, running nearly at right angles with the main gangway. The same bed of coal may be pierced by several gangways—diverging from each other as the avenues diverge from the Capitol at Washington—from each of which extend numerous lateral chambers. These chambers often intersect each other, and thus constitute a network of passages like the streets of a city. Along the principal passages tramrails are laid for the transportation of the coal in trams, or little cars, from the remote portions of the mine to the shaft. Each miner employs a separate tram, and receives a stipulated amount per ton for the coal sent up by him. The trams are moved over the track by mules, which often spend their lives under ground. They

Fig. 62. Explosion in a Coal Mine.

are stalled and fed in side-rooms excavated in the coal and superincumbent rocks. The requisite circulation of pure air is maintained through the mine by the consumption of refuse coal at some suitable place, the smoke and heated air from which ascend through a separate shaft. The escape of heated air through this shaft causes a descent of external air to take place through the main shaft. Communication between the two shafts is effected only through the remote portions of the mine, so that the pure air is made to permeate all the passages. Still there must always be side-rooms through which no circulation can be effected, and here not unfrequently collects that explosive "fire-damp," or light carbide of hydrogen, so often evolved spontaneously from the coal, and so often the cause of fatal accidents to the miners (Fig. 62). When the seam of coal is less than five feet thick, it becomes necessary to remove some of the superincumbent rock, to render the roofs of the main passages sufficiently high for the mules to travel under them.

Thus entire square miles of a coal-seam, hundreds of feet beneath the surface, are perforated in all directions by the hand of the miner (Fig. 63), as ship-timber is riddled by the deprĕdations of the *Teredo.*

By the feeble light of our miner's lamp we enter one of these dusky aisles. The substratum beneath our feet has been ground to dust. The whole thickness of the coal-seam is exposed along the lateral walls. Occasionally it presents gentle undulations instead of lying in a rigidly plane position, and not unfrequently a huge bulge of the underlying rocks completely cuts off the seam. Overhead a black, bituminous shale forms the ceiling. Perhaps here and there the white shell of a univalve or a bivalve projects from the surface—the products of the sea buried in their native sediments, and suspended above our heads.

Fig. 63. Miner at Work—old manner of working.

What a change in the condition of things since those little animals lived in the shallow surface-waters in which those sediments accumulated! Lo! here above us is a mirror-surface gleaming in the light reflected from our lamps. Its polish is like that of jet, and yet it is wrought upon the face of the solid rock. Some slight movement of the earth's crust has cracked the shaly roof; the opposite sides of the fissure have been moved to and fro over each other, and under the mighty pressure the two opposing faces have been beautifully polished.

But probably different sights will greet our eyes. The rocky ceiling is ornamented every where by the most exquisite tracery—inimitable representations of the delicate fronds of ferns (Fig. 64). We remove a scale of the rock, and behind is still another picture. Remove a second, and from the dark black rock gleams forth another form of grace and beauty. The whole mass of the shaly roof is a portfolio of inimitable sketches. The sharpest outlines and minutest serratures of the leaves are clearly traced. The very nerves, with their characteristic bifurcations, are accurately depicted on this wonderful lithograph. Petioles, and buds, and woody stems, and cones, and fruits, slender grass-leaves, striated rushes, the

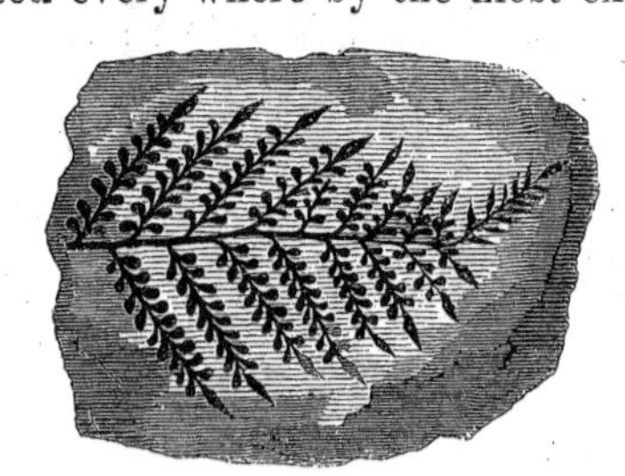

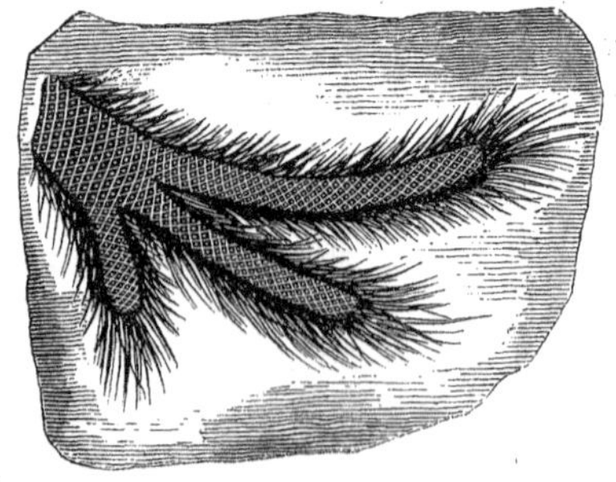

Fig. 64. Coal Ferns.

fluted stems of gigantic club-mosses, the scarred and pitted trunks of extinct tree-ferns, diversify, by turns, the crayon sketchings of the dusky ceiling. Prostrate, all! They have stood erect; the soil has held them by their spreading roots, the genial sunlight has warmed them, the vital breeze has fanned their verdant foliage; change, which transforms all things, has swept over them, and graceful fern and giant club-moss, slender reed and arrogant conifer, have laid down together in their couch of sediment, and the old sexton, Time, has piled upon them the accumulated ashes of a hundred succeeding generations of trees, and herbs, and perished populations. What a store-house of suggestions is here! The dusty "Catacombs" are less eloquent in their inscriptions; the vaults of the Pyramids recite a history less full of meaning. To the soul that holds communion with the visible ideas that dwell about him, these rocky walls are vocal with narratives of earthquake and flood, of nodding verdure and of desolating surge; these shales are the tombstones of generations, on which are inscribed chronologies whose minutes are the cycles of the Hindoo. Here is the populous abode of world-ideas. Through these dim avenues flit spectres of the ancient thoughts which were once the acting energies of our planet. Here is the real Acherontian realm. He who has descended to these subterranean halls, and held converse with the forms which here abide, has visited a world and communed with intelligences of which Anchisiades had only dreamed.

Shall we venture to translate the histories recorded upon these rocky leaves? What were the scenes and events of that epoch of the world when these buried vegetable forms were living, growing organisms, and Nature was storing away for the human race these magazines of fuel?

CHAPTER XIV.

THE SCENERY OF THE COAL PERIOD.

IT was in the middle ages of the history of the world. The growing continents had lifted their brows above the surface of the all-embracing sea; but their spreading plains and long-extended shores were still the empire of the garpikes, and the nursery of illimitable beds of encrinites and polyps. The Gulf of Mexico jutted northward to Middle Iowa, and rolled its widening waters northwest far toward the sources of the Missouri River. There are good reasons for believing that it stretched through the entire length of the continent to the Frozen Ocean. The shore-line of the Atlantic reached from Connecticut through Southern New York and Northern Ohio, Indiana, and Illinois, to the valley of the future Mississippi. All the centre of Michigan was a sea-bottom, and not unlikely a gulf projected northward over the peninsula now inclosed by the great lakes. There was never, however, any free communication between the Michigan Gulf and the ocean after the later portion of the Devonian Age. Hudson's Bay stretched far toward the site of Lake Superior, as the Arctic Sea pushed down from the north to fall into the warm embrace of the waters of the Mexican Gulf. The great lakes were not—save, perhaps, Lake Superior—nor the mighty Mississippi, nor the thunder-voiced Niagara. The youthful continent was yet unclothed with soil, save the rocky detritus which nourished the lean vegetation which began to garnish the land during the period of the Chemung and Marshall. The skeleton rocks protruded every

Fig. 65. Ideal Landscape of the Coal Period.

where in bleak, inhospitable exposures. Occasionally in a low valley was gathered a cluster of dwarfish trees, nourished by the crude aliment of a hastily-compounded soil. Beast, and bird, and insect were yet slumbering in the chambers of the future—ideas reserved in the all-producing mind of Omniscience. Food for them there was none. The atmosphere was a noxious poison, charged with all the carbon which now exists in the form of modern vegetation and beds of mineral coal. Denizens of the sea had for ages strewn its bottom with the ruins of their workmanship—mountains of coral masonry had been reared by the little polyp architect, but in all the murky air which floated over the land and sea was not one motion of an animated being—not a voice—no song of bird, or hum of insect's wing to break the dread, eternal silence. The surges broke upon the beach, the tempest gathered in the thickening air, but no beast hurried to the sheltering cave; the storm burst upon the bald and desolate cliff, but no fluttering wing sought protection from its fury.

The period had now arrived, however, when this verdureless and voiceless scene was to be clothed and animated. Now was perhaps the most important epoch in the whole physical history of our planet. The forces of nature were now to be called to their grandest exercise. The laws of chemistry were summoned to an operation miraculously beneficent and providential. Organic force now girded itself for the production of new and higher forms of animalization, and for the display of the earliest and richest exuberance of the vegetable kingdom.

The series of animate existences began with the protozoön, and had been carried through long progressive stages to the highest types which make their home in the water and respire that element. Man, the far-off consummation of all these improvements, was to be a vastly superior being; but the next step in the direction of this consummation must be the introduction of an air-breathing animal. In the existing condition of the world no air-breathing animal could survive, and Nature was called upon to solve the problem of the elimination of the noxious gas which unfitted the atmosphere for respiration. Till this was done the progressive series of animal forms must here be arrested, and the last term of the series, man, toward which all the steps of the previous preparation had converged, must remain a distant and unattainable hope, and Nature fail of her completeness and her crown.

The development history of the American continent had been conducted through a succession of vertical oscillations, extending eastward to the still subaqueous ridges of the Appalachians, and westward to the corresponding nascent ridges of the Pacific slope. The valleys of the two great oceans had been continually deepening beneath the pressure of the superincumbent masses of waters, and, as a consequence, the intervening continental space had suffered a corresponding vertical uplift, so that the waters had been poured off from the site of the future continent, and a mere shallow lagoon occupied the present area of the Middle and Southern States and Territories. The oscillations of the submarine soil down to the dawn of the period now under consideration—sometimes increasing and sometimes diminishing the depth of the waters—left it at last but little sunken beneath the general surface of the sea. [See the areas marked C, Fig. 58.]

Now a state of more than usual uneasiness began to manifest itself. The ocean bed heaved and sank as in the breathings of a mortal agony. Surges mountain-high rolled up the sterile strand, and, wasted with their own violence, fell back upon their ocean couch. This, of course, was not the period for an abundance of animal life. But, if the usual fecundity of Nature was for a time suspended on our continent, some other continent may have been the theatre of its display. In America the crumbling margins of the sea were worked up into cubic miles of sand and pebbles, and transported to embankments sometimes thousands of feet in thickness. The tombs of the *Cephalaspis*, *Pterichthys*, *Coccosteus*, and *Holoptychius* of the Old World, and of *Onychodus*, *Machæracanthus*, *Agassichthys*, and *Rinodus* of the New, were buried immemorable depths beneath the rubbish of a geological revolution. These were the accumulations of the "Millstone Grit." Anon, the violence of Nature suffered a pause, and finer sediments only were transported over the areas previously strewn with sand and pebbles. Many alternations of finer and coarser deposits thus succeed each other among the lower beds reposing immediately beneath the coal.

In the course of ages the shallow sea became a marsh. Now that a foothold for terrestrial vegetation was established, the all-adaptive hand of Nature planted the soil with many kinds of herbs and trees. Simultaneously, on every side, innumerable germs spring up from the new-made sediments. Vegetation, in varied types and family alliances, starts forth at the fiat of creative energy, and the world is dressed in a garment of shining verdure. No provident hand had strewn the soil with the seeds of these unfamiliar forms. The All-commanding had summoned the tribes of plants from the shadowy realm of ideas, and they stood forth in multitudinous array, clad in the newest and brightest garments of Nature's exhaustless wardrobe —mute, unconscious existences, but yet with life and organs, beginning from the moment of their appearance to play upon the elements which Omniscience had provided for their elaboration. How carefully was the soil prepared to encourage the luxuriant growth and wide dissemination of these beautiful creatures! Lifted above the level of the sea, it maintained the humid condition most congenial to the nature of the most luxuriant growers. The internal heat of the earth, however, at this early period, warmed the surface to a tropical temperature, and stimulated the roots of the new-born vegetation, while from the tepid waters the atmosphere was reeking with moisture, and ever and anon dispensing its showers upon the green-carpeted savanna. But, more than all, the food most grateful to the growing plant was that abundant carbonic acid whose presence in the atmosphere was the fatal bar to the introduction of terrestrial animals.

This scene of verdure was destined to short duration. One of the ever-recurring oscillations of the earth's crust sank the entire flora beneath the ocean's level. Pebbles, and sand, and argillaceous mud were strewn over the layer of prostrate vegetation, and the sea again held undisputed sway over states once rescued from its dominion.

Again the established order of Nature brought these latest sediments to the surface, and again, as if by magic, the fairy forms of a flowerless vegetation start up from the germless sands. Generations of these new forms luxuriate in the humid vales of another epoch—fix, in their woody tissues, another portion of the superabundant plant-food of the atmosphere, and then fall down to mingle with the peaty accumulations of the period.

Anon, another inundation devastates the scene, and sands and clays are borne by the rushing tides, and the dense growths of the recent jungle again disappear beneath another packing of silt and shingle, as a field of marsh-grass is buried beneath the sand borne forward by the summer overflow of a great river. Thus, perhaps, a hundred times in the course of ages, the vegetable growths of one epoch were entombed beneath the *débris* of a more violent one. Occasionally the inundating waters assumed the quiet habit of a deep and permanent sea. Then, that no adaptation of inorganic nature might be wanting in the answering aptitudes of the organic world, myriads of marine creatures swarmed, and lived, and died upon the grounds that had often aforetime been the seat of terrestrial vegetation. Thus, perhaps, a bed of calcareous sediments, destined to become a limestone, was interpolated among the couches of sand, and shale, and vegetable matter.

The theatre of these changing scenes was the whole of that area now covered by the coal-measures of the country (see Fig. 57), as well as large portions of the intervening regions, from which the coal has been swept by the besom of geological denudation. In the later ages of geological history, wasting agencies have moved over the surface of the country, scoring through the solid rocks, scooping out lake-basins, carrying away entire formations, and exposing deeply-seated strata over wide areas.

The duration of the vicissitudes which I have sketched was inconceivably great. The amount of vegetable matter in a single coal-seam six inches thick is greater than the most luxuriant vegetation of the present day would furnish in twelve hundred years. Boussingault calculates that luxuriant vegetation at the present day takes from the atmosphere about half a ton of carbon per acre annually, or fifty tons per acre in a century. Fifty tons of stone-coal, spread evenly over an acre of surface, would make a layer of less than one third of an inch. But suppose it to be half an inch; then the time required for the accumulation of a seam of coal three feet thick—the thinnest which can be worked to advantage—would be seven thousand two hundred years. If the aggregate thickness of all the seams of coal in any basin amounts to sixty feet, the time required for its accumulation would be one hundred and forty-four thousand years. In the coal-measures of Nova Scotia are seventy-six seams of coal, of which one is twenty-two feet thick, and another thirty-seven. The

"Mammoth Vein," at Wilkesbarre, Pennsylvania, is twenty-nine feet thick. Add to the time occupied in the accumulation of the coal the time which elapsed during the inundations, when shales, sandstones, and limestones accumulated to an average of fifty times the thickness of the coal, and we shall have at least double the above interval, or two hundred and eighty-eight thousand years, for the time required to build up a series of coal-measures three thousand feet thick. This is about the thickness of the Pennsylvania measures, while those of Nova Scotia are five times as thick.

It is not forgotten that allowance must be made for the extraordinary luxuriance of Carboniferous vegetation, and I offset this consideration by the fact that large quantities of carbon, taken by it from the atmosphere, must have been returned again by the partial decay and destruction of the tissues, thus rendering them so difficult to detect in the substance of the coal. Other calculations, based upon the assumption that the coal-measures were accumulated at the mouths of rivers, result in the determination of a length of peroid equally enormous. But the whole history of our world since the commencement of animal existence is divided into over thirty periods, each corresponding to that of the coal, and that portion of its history anterior to the creation of animals was at least equally protracted!

The vegetation of this period was comparatively low in rank. It was almost exclusively a flowerless vegetation. But the sombre aspect of the prairie and forest comported well with the absence of admiring intelligences, and the low grade and character of the few beings which basked in the sun or bathed in the waters of the Carboniferous Age. The leading forms of vegetation were allied to *rushes*, *ferns*, and *club-mosses* (see Fig. 65). Many of these grew to colossal dimensions. Some of the rushes—*Cala-*

mites—were doubtless thirty feet in height. The impressions of their huge and prostrate stems may often be traced upon the shale which overlies a seam of coal. Of ferns, no species living in temperate latitudes attains the dimensions of a tree; but there formerly flourished within the limits of the Northern States ferns which attained to arboreal dimensions, single fronds of which reached the length of six to eight feet. The club-mosses—*Lepidodendra*—of the same epoch grew to the magnitude and aspect of stately palms (Fig. 66). Among us they trail upon the ground, or rise but a few inches above it. The largest living club-

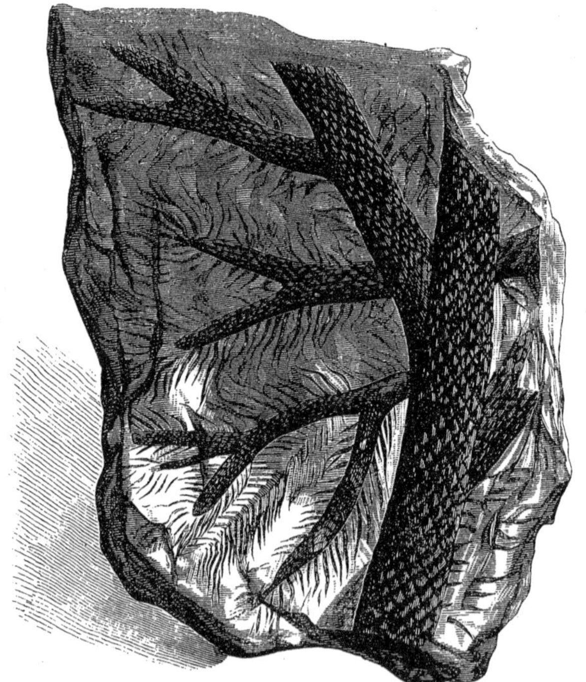

Fig. 66. Remains of a Lepidodendron.

mosses do not exceed three feet in height. The stems of *Lepidodendron*, after the falling of the leaves, were covered with scars diagonally arranged, and are often mistaken by the uninformed for "petrified snakes." The cones of these plants are found in great abundance in Ohio. Another curious form of this period has been styled *Sigillaria*. Their fluted trunks, from one to five feet in diameter, have sometimes been seen sixty and seventy feet in length. The flutings are marked by a longitudinal series of pits, like the impressions of a *seal*. In many instances these tree-trunks have been found erect, evidently buried while standing by accumulations of sand and mud (Fig. 67). Below are the roots and rootlets—formerly called *Stigmaria*—and the very soil remaining in which they flourished. In the excavation of a bed of coal these petrified tree-trunks are not unfrequently cut off below, when the slight taper of the stem permits them to slide, by the force of gravity, down into the mine. These "coal-pipes" are much dreaded by the English miners, for almost every year they are the cause of fatal accidents. "It is strange to reflect," says Sir Charles Lyell, "how many thousands of these trees fell originally in their native forests, in obedience to the law of gravity, and how the few which continue to stand erect, obeying, after myriads of ages, the same force, are cast down to immolate their human victims."

Let the reader embody before his mind's eye a group of rush-like and fern-like trees and under-shrubs, interspersed among gigantic club-mosses and occasional conifers, and he has a picture of a carboniferous jungle—a jungle not enlivened by the tread of quadrupeds or the singing of birds, but mute as the solitudes of an African desert—voiceless save when the alligator-like bellowings of the *Archegosaurus* in a neighboring bayou waked the echoes of those gloomy corridors, and startled the lesser amphibia from

Fig. 67. Trunks of Sigillaria in the Mine of Treuil, at Saint Etienne.

their hiding-places, or the thunder-voice of Deity spoke, as it still speaks, from the terror-striking tempest.

The office of this redundant vegetation was finally fulfilled. The atmosphere was purified of its noxious elements, and higher creatures could live upon the soil. Behold the wisdom and providence of the creative Architect! Carbonic acid was to be removed from the atmosphere, to fit it for animal respiration. A finite mind might have aimed to effect this end alone. Omnipotence was competent to annihilate the poison, or convert it to some solid or liquid form. The infinite Intelligence, however, had so planned the universe that the poison of the quadruped was the food of the plant. The very execution of one portion of the cosmical plan created a use for that which impeded the execution of another. A double object was thus effected. Nor was this all. Should these enormous crops of vegetation grow up and pass away unutilized for the want of an intelligent population to consume and use the fuel? It was not so to be. Though man was not except in the conceptions of the Almighty, man was regarded in the preparations of this age. The far-seeing Planner of the universe stored the carboniferous fuel in repositories where it could never perish, and where it could await the uses of the coming race of man. Nor was this even the end of the providential purposes. In a subsequent age those barren rocks and those beds of coal became covered, first with the basis of a soil, and then with the soil itself; so that man, when he should come upon the stage, might find an inexhaustible mine of fuel, and a foothold for the products of his farm, upon the self-same acres. Another circumstance should also be here remarked. The preservation of these carboniferous stores was effected by the packing down of layer after layer, while beds of clay, and sand, and calcareous sediment were interposed between them. Not only was there never another period of the world when the supply of carbon was so great, but never, before or after, were those frequent and gentle oscillations so long continued which were the agencies in burying the successive crops of vegetable growths. At least, such frequent oscillations never before or since occurred at a time when the level of the continents so nearly coincided with the level of the sea. And, lastly, these very oscillations, while they were subserving this collateral end—which was still important enough to have been the sole and ultimate end—were only the symptoms of a great continental preparation, which was going on from the region of the Atlantic to that of the Pacific shores, and which had been in progress, and attended by similar, though much less frequent oscillations, from that remote period when the shrinking of the molten nucleus of the world located those huge wrinkles in the stiffening crust which were to be afterward deepened into the beds of the two great oceans. Verily, here is a scope and comprehensiveness of plan which must command our highest admiration.

The same general preparatory movements were still to be continued—continued till the finished earth had been elaborated for the reception of man. It would seem that the frequent oscillations of the Coal Period were but the tremblings of the strained crust, pushed to the very verge of violent rupture by the two enormous masses of water. By turns, the central areas had been protruded above the waves, and by turns the tension had found relief, and the uplifted crust dropped back for a time to its submarine horizon. Not before the collateral uses of these phenomena had been subserved did the tension of the crust reach the measure of a grand upheaval. After trembling for ages beneath the immense and increasing pressures of two great oceans, it burst up in enormous folds thousands of feet in height, and extending from New England to Alabama. Some of the folds of the Appalachian upheaval, according to the grand generalizations of the brothers Rogers, were protruded with so abrupt a flexure, and to such a dizzy height, that they toppled over toward the west; while to the west of the principal axis of violence the folds become gentler, and terminate in pleasant undulations of the surface. The Queen City of the West stands, perhaps, on the last of this series of undulations. Thus were the Appalachians brought into existence. Thus were the deep-seated beds of coal lifted above the general level of the land, and brought within the reach of moderate excavations, accompanied by the requisite conditions for natural drainage of the mines (Fig. 68).

Fig. 68. The North American Continent at the end of Paleozoic Time, or beginning of Mesozoic. (The dotted lines represent its present outlines; the broken lines the rivers.)

Subsequent geological agencies have greatly modified the primary result. The ocean has been permitted still again to sweep over the continent, and the crests of the folds and ridges have all been planed down, and the materials distributed over the intervening spaces, or worked up in the rock-building of later ages. Thus the original height of the Alleghanies has been much reduced. Thus the swell upon which the Queen City of the West is built has been worn off to the level of the adjacent areas; and thus the original limits of the great Carboniferous jungle have been very much restricted.

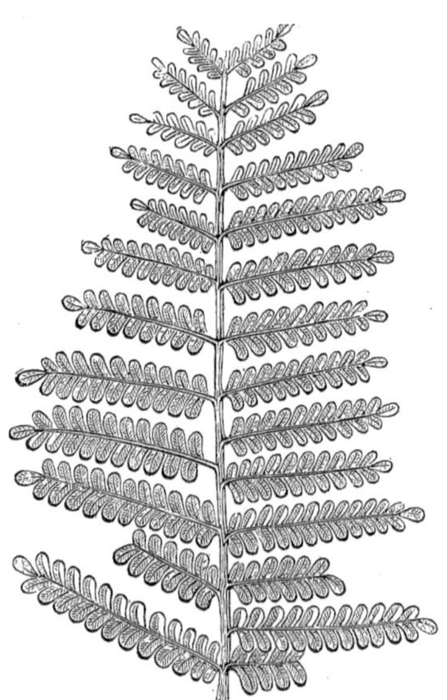

Pecopteris. Frond of an ancient Tree-fern.

CHAPTER XV.

THE SCOUTS OF THE REPTILE HORDE.

EMPIRES rose upon the earth, and crumbled in succession to decay, a thousand ages before the foot of Adam had pressed the soil of the Garden of Eden. A series of dynasties flitted like shadows over the face of our planet, and disappeared beneath the dim horizon of the past, while the empire of man was but an idea dwelling in the Almighty Mind. Here were morning and evening, invigorating sunlight and cooling dew, softly-wooing breeze and fiercely-maddened tempest, springtime and autumn, weeping clouds and placid evening sky, Winter piping his melancholy song upon the withered reeds of Summer, ocean-surges waging everlasting battle with the rocky shore, God alone spectator of the progress of the mighty work which was being accomplished. But there was life, and motion, and consciousness, and enjoyment, and death through all those dim and distant ages. Those dim and distant ages—how imagination halts, and faints, and falters in the effort to shoot back over the infinite stretch of years! Life was here, but without a voice, without a wing, without a footstep. The ignoble mollusc held dominion in the sea through all the morning twilight of animated existence.

The mute fish reared his empire on the ruins of that of the mollusc. In the middle Paleozoic ages this first and lowest form of vertebrate existence appeared in all the seas—not fishes clothed in horny scales like those which swarm in the waters of the human era, but fishes clad in coat of mail, bucklered aud helmeted with bony plates, and armed with long and powerful spines, or, in a later age, with a fearful array of sharp and conical teeth. The dynasty of the fishes sprang up in that period when the limestones of Buffalo, in New York, and of Columbus, Sandusky, and Kelly's Island, in Ohio, were accumulating as sediments in the bottom of the sea; when Canada West was the ocean's bed, and the last crop of zoöphytes was growing upon it; when the beautiful island of Mackinac was a submarine plantàtion, and the embryo fastnesses of Old Fort Mackinac witnessed an onslaught and a massacre more bloody and destructive than that of 1761. The empire of the fishes waxed more powerful during the succeeding epochs, when the "black shales" of the West, and, later, the beautiful sandstones of Waverly and Cleveland, Ohio, were the ocean's bed, and hordes of marine forms roamed over the area of Southern New York, and nearly the whole of Michigan, Indiana, and Illinois. The Marshall epoch probably covers the latter part of the period of the Old Red Sandstone of Scotland, whose ichthyic populations have been so graphically described by the author of the "Asterolepis of Stromness."

The reign of the fishes was prolonged through the Carboniferous period; but the types which wielded the sceptre during the later ages of the empire assumed less questionable forms, and began to approach the external configuration of the fishes of our day. They were mostly clothed, however, with bony scales, and the backbone extended into the upper lobe of the tail, which was longer than the lower; or, what is probably a more correct view of this structure, the tail was supplied upon the under side with a supernumerary fin, the development of which deflected upward the true caudal fin—the tail of the sturgeon and the garpike being as truly "homocercal" as that of the whitefish. It is sad to think of the ancient populousness and prowess of these mail-clad fishes, and then turn to our own times and find them reduced to a few isolated, hated, and hunted species. The garpike or "billfish" (*Lepidosteus*), and the sturgeon (*Acipenser*), are the only surviving representatives of the royal families of the Carboniferous Age. In turn, the dynasty of the fishes was superseded by that of the reptiles.

It was impossible that air-breathers should inhabit the earth before the atmosphere became purified of the noxious gases which remained from the ancient igneous condition of the globe. The principal impurity—carbonic acid—was destined to be consumed by the demands of an abundant terrestrial vegetation. The latter part of the reign of fishes was marked by the advent of multitudes of land-loving vegetable forms—the heralds of the close of the dominion of races whose element was the water. It was many ages after its first appearance before terrestrial vegetation became fully established. We know that here and there one of these stranger forms grew upon the shores of those seas which were the domain of the fish; and, falling down upon the beach, or borne along by river torrents, the decaying trunks were drifted seaward, and sunken among the sands which entombed the bodies of the royal families of the age. We know that the slight improvement in the condition of the atmosphere was responded to by the introduction of a few air-breathers of sluggish and imperfect respiration. The name of the oldest air-breathing animal at present known to have lived upon our earth is *Telerpeton Elginense.* Its remains have been found in the south of Scotland, in a yellow sandstone supposed to be of the same age as the Old Red Sandstone. The same rock has furnished some other remains, formerly supposed to be the vestiges of fishes, but now known to be the remains of reptiles; and geologists are not by any means of one accord in the opinion that those sandstones are older than the Carboniferous Age, or even older than the Trias. Besides these, the most ancient traces of reptilian remains occur in the coal-measures, which were deposited during the decline of the empire of fishes in the latter part of the Carboniferous Age.

The geological history of reptiles possesses many points of extreme interest; and, in order to make them clear to the reader, and to give precision to the brief account which I am about to furnish, I shall endeavor to recall in few words the classification of this group of vertebrates.

Reptiles proper, in point of rank, are next above the Batrachians, which come next above the Fishes. Reptiles are purely aerial in their respiration; Fishes purely aquatic; while the Batrachians breathe water in infancy, and air at maturity, exhibiting thus a compromise between the ichthyic and reptilian modes of respiration. The body of the reptile is always covered with scales or bony plates, while that of all modern batrachians is smooth or "naked." The vertebræ of most reptiles are concave at one extremity—generally the anterior—and convex at the other; the vertebræ of batrachians are concave at both extremities, like those of fishes. There are other distinctions to which I need not refer. The frog is the type of the highest order of existing batrachians, the salamander of the second, and the "fish-lizard" of the lowest. The first is possessed of a tail only in the young or tadpole state; the second retains its tail during life; and the third retains both its tail and aquatic—or embryonic—mode of respiration.

Of reptiles, three orders which have played a most conspicuous and important *rôle* in the history of the world are entirely extinct, and three others still survive. The turtles, saurians, and serpents, in descending order, embrace existing reptiles. The first are inclosed in a carapace or "shell;" the second have elongated forms, generally clothed with scales or bony plates, and almost always possess four extremities; their eyelids are movable, and their two jaws move vertically like those of higher animals. The serpents are equally clothed with scales, but their bodies are more elongated, and destitute of limbs; their eyelids are immovable, and each of their jaws is in two pieces; and they have, besides, an extra pair of jaw-pieces in the roof of the mouth. The extinct orders are Pterodactyls, or flying reptiles, Enaliosaurs, or marine reptiles, and Labyrinthodonts, or reptiles with very complicated structure in the substance of their teeth, and, sometimes at least, with frog-like forms. They possess affinities both with true reptiles and with batrachians.

In 1828, Dr. Duncan, a Scotchman, had his attention arrested by what appeared to be tracks of a reptile imprinted upon the surface of solid sandstone at Dumfriesshire. A few years later, tracks somewhat resembling the impression of a human hand were observed upon similar sandstone in Saxony. These were also attributed to reptiles. In this country, Dr. Deane and Professor Hitchcock noticed upon red sandstones, in the valley of the Connecticut River, numerous tracks which they were inclined to attribute to birds, as they were evidently made by *three-toed bipeds.* In 1836, Professor Hitchcock published the first systematic account of these footprints, in which he pronounced them to be mainly the tracks of birds—*Ornithichnites*—a conclusion which is very questionable. In 1844, Dr. King, of Philadelphia, also described several kinds of footprints upon rocks then supposed to be carboniferous, but since shown to be of the same age as the sandstones upon which all the other known tracks had been observed. The rocks are the "New Red Sandstone," belonging to the lower part of the Jurassic system, or the upper part of the Triassic. The position is a considerable distance above the coal.

The first indication of the existence of reptilian remains in rocks as old as the coal was the discovery, in 1843, by Sir William Logan, of some footprints in the coal-measures of Nova Scotia. The first reptilian bones were discovered in 1852, in the celebrated coal-measures of South Joggins, on the Bay of Fundy. The measures here are two and three fourth miles in thickness; and along a middle belt of fourteen hundred feet they abound in the remains of ancient forests, the trunks and stumps of large trees still standing erect, with their roots still penetrating the ancient soil. Here, as has been shown by Messrs. Dawson and Lyell, root-bearing soils occur at *sixty-eight different levels*, and between them are deposits of shale and sandstone, which must have had an aqueous and probably a marine origin, thus showing, beyond all controversy, that the level of the locality underwent at least sixty-eight oscillations during about *one tenth* of the period of the coal-measures. Many of these fossil tree-trunks are hollow, and filled with sandstone containing vegetable remains. In one of these hollow trunks the hammer of the Acadian naturalists laid bare some bones, which proved to be the remains of the oldest reptile at that time known in America, and which was subsequently named *Dendrerpeton Acadianum*. Different individuals must have varied from six inches to three feet in length, and they were probably batrachians rather than true reptiles, though naturalists do not always make the distinction. These little animals seem to have made their home in the hollow of the tree, and to have been overtaken by the flood which ended the epoch and buried them among the other relics of their time. Another batrachian was discovered the same year in the coal of Pictou, in Nova Scotia, and in 1859 still another. The reader will find these all more minutely described in Dawson's "Air-breathers of the Coal Period."

In 1856 the first Batrachian bones were described from the United States. These were discovered by Dr. Newberry and C. M. Wheatley, at Linton, Jefferson County, Ohio. There were three different types of beings. The first had the head and ribless trunk of a frog, combined with the limbs and tail of a salamander. The second and third had the vertebræ of a salamander, with the ribs of a serpent. The first of these animals has been named *Raniceps Lyelli* by Dr. Wyman, of Boston.

In 1863, Professor O. C. Marsh described, from the coal-measures of South Joggins, Nova Scotia, the remains of a reptile somewhat higher in rank than any other previously known in rocks of so high antiquity—a true reptile belonging to the Enaliosaurs, or marine saurians, and related to the huge reptiles which sported in the waters of the Mesozoic time, some of which have been so genially described by Dr. Mantell. This animal, which is believed to have been from twelve to fifteen feet in length, was probably one of the most fish-like of Enaliosaurs. It has been named *Eösaurus Acadianus*.

How scattered must have been the air-breathing population of the globe when, after thirty years of careful observations, geologists have brought to light only the foregoing brief list from the carboniferous rocks of the country. I make no note of two or three species of air-breathing snails, a myriapod, and two or three orthopterous insects. Seven species only of vertebrate air-breathers—intrepid forerunners of the numerous populations of the succeeding periods—scouts, sent forward upon the earth to spy out the land, and test its fitness for the occupancy of the hordes which were to follow!

The coal had been deposited; cubic miles of fuel for the consumption of future generations had been taken from the atmosphere, and packed in beds of clay and sand, to await the arrival of a far-off race. The air was fit for the respiration of a low order of terrestrial animals, and, in obedience to the mandates of creative energy, they began to come forth. There was an interval of time in the history of the world when the scales of empire hung balanced between the fishes and the reptiles. The first were on the wane; the latter were gathering strength from age to age. Nature favored the latter. Omnipotence bade them march on, and vanquish and sweep from the earth those lower forms which had been permitted to hold the mastery in creation only because the world was, as yet, unfit for beings more exalted and worthy. These middle ages are styled the Permian period. Vegetation was still abundant. Though the acme of vegetable luxuriance had passed, and no more vast deposits of coal were to be treasured up—at least in those portions of the world preparing for the occupancy of the Caucasian race—the trunks and leaves of the flora of that period, preserved in beds of sandstone and shale, attest the productiveness of the Permian soil. In these ages of the world the first emphatically lizard-like reptiles came upon the stage—a family belonging to the Saurians—though many of the Permian reptiles present a divergence from true lizards in having their numerous teeth implanted in sockets instead of soldered to the margins of the jaws. Occasionally, also, was to be seen the frog-like form of a Labyrinthodont sunning himself upon the marshy border of a Permian estuary (Fig. 69).

Until within a few years geologists were unacquainted with any Permian rocks upon the American continent. Dr. Emmons, however, in 1856, established the existence of Permian reptiles in a brown sandstone in North Carolina, deposited in some of the furrows between the folds of the Appalachian upheaval. A little later, Permian rocks were announced from Kansas almost simultaneously by

Swallow, Meek, and Hall. The Permian rocks west of the Mississippi have more recently been subjected to thorough investigations by Swallow, Meek, Hayden, and Shumard; and though much of what was first regarded as Permian is now proven to belong to the period of the coal-measures, we know that this group of sediments is developed in America on a scale little less magnificent than in the ancient Russian kingdom of Perm, which gave its name to the group.

Fig. 69. Labyrinthodon (restored).

CHAPTER XVI.

THE REIGN OF REPTILES.

WE now enter upon a new Age in the history of the world. The reptilian army has arrived and taken absolute possession of the land and the sea. The highest of these monarch reptiles, in respect to organization, are some lizard-like creatures as much at home in the sea as upon the land. But especially does the great deep swarm with huge beings having the body of a lizard, the immense jaws and sharp, conical teeth of a crocodile, the bi-concave vertebræ of a fish, and the short, flat paddles of a whale. What a synthesis of characters is here! The type of the age is expressed in the lacertiform trunk and tail. Crocodiles had not yet existed; but the jaws of these monsters seem like an experiment preparatory to the supply of quantities of those savage brutes. Mammals were yet in the distant future; but here, in the paddles of these Enaliosaurs is a prophecy of coming cetacea—the form which the mammalian type assumes at the point where it comes in contact with the type of fishes. The reign of fishes is past; but here, in the bi-concave vertebræ of these sea-monsters is preserved a reminiscence of the last sovereigns.

This, the Triassic Age, was peculiarly the reign of Labyrinthodont saurians. Thirty-five years ago the tracks of these anomalous creatures were first noticed upon some red sandstone in Saxony, and they have since been discovered in other parts of the world. The peculiarity of these footprints consists in their hand-like form, and in the occurrence of a series of larger and smaller in connection with each other. The latter circumstance led to the opinion that the posterior limbs of the reptile were much stouter than the anterior, as in the kangaroo and frog. When the bones of these animals were brought to light, geologists had the opportunity to certify themselves that these problematical hand-prints were impressed by reptilian instead of mammalian quadrupeds; and that while the weight of characters allied them to true reptiles, they nevertheless possessed strong analogies with Batrachians, and probably simulated the form and habits of the frog—though in truth we should say that the frog was subsequently fashioned in the similitude of a Labyrinthodont. The head was helmeted by a pair of broad, bony plates, through which were openings for the eyes; and some parts of the body were covered, especially in the later ages, by a similar armor. The striking characteristic of these ancient reptiles, from which they receive their name, is seen when a very thin transverse section or slice of one of the teeth is viewed under the microscope. The external coating of the tooth, called cement, is folded inward in folds which reach to the central cavity, and in their course are inflected into a *labyrinth* of subordinate lateral folds. Some of these frog-like quadrupeds seem to have attained the size of an ox. It is likely that they were the representatives of the class of Batrachians in those early periods, as no other Batrachia are known in the Trias; and those before alluded to from the coal-measures are known likewise to have possessed the peculiar cephalic plates of the Labyrinthodonts.

The Triassic Age witnessed also the advent of multitudes of marine saurians of the family of Ichthyosaurs, having enormous cavities in their craniums for the lodgment of the eyes. This type of reptiles is restricted to this single age of the world. Here also crawled reptiles resembling gigantic lizards, semi-aquatic or purely terrestrial in their habits, having feet for walking instead of flat, oar-like extremities for swimming.

These forms all disappeared with the dawn of a new era. Their bones lie buried in the geological cemeteries of Europe. It is almost incredible that information so exact can be drawn from the few scattered fragments which have been brought to light; but such is the unity and persistence of plan which runs through the different classes of the animal kingdom, that a single tooth, whether of a living or extinct species, will often suffice to enable the expert to disclose all the zoölogical relationships of the animal to which it belonged, to delineate its form, and size, and habits of life; as the architect from a single capital rescued from a ruined edifice can declare not only the general style of the entire architecture, but can reproduce the size and proportions of the temple whose spirit and method it embodies. Not less sublime than the work of the astronomer, who sits in his observatory, and, by the use of a few figures, determines the existence and position in space of some far-off, unknown orb, is that of the paleontologist—the astronomer of time-worlds—who, from the tooth of a reptile, or the bony scale of a fish found thirty feet deep in the solid rock, declares the existence, ages ago, of an animal form which human eyes never beheld—a form that passed totally out of being uncounted centuries before the first intelligent creature was placed upon our planet—and by laws as unerring and uniform as those of the mathematics, proceeds to give us the length and breadth of the extinct form; to tell us whether it lived upon dry land, in marshes, or in the sea; whether a breather of air or water, and whether subsisting upon vegetable or animal food. It is this unity of the laws of animal life and organization running through the whole chain of existence, whether past or present, whether extinct or recent, that constitutes the sublime philosophy of paleontological studies, and assures us that one enduring and infinite Intelligence has planned and executed every part of creation.

Crowds of reptile forms have passed before our view, but we have only just arrived at the culmination of the reign of reptiles—the *Herpetarchy* of the world's history. The Jurassic Age followed the Triassic. Before this time the Trilobites of the Paleozoic Ages were known only in history. The plain-chambered shells had been followed by lobulate-chambered shells—the *Goniatites*—and these were now, to a great extent, superseded by the *Ammonites*, a family of chambered shells with dorsal siphons and extremely complicated partitions between the chambers. So the complexity of Nature's products increased with her age. Most of the Ammonites were closely coiled. In their modifications and decorations the exuberance of Nature effloresced in hundreds of distinct species. Six hundred representatives of this peculiarly European family are exhibited in the museum of the University of Michigan—one of the results of the tireless industry of Dr. C. Rominger. The land was clothed with a vegetation quite similar to that of the present day; but the climate was yet warmer than at present, and many types of plants and animals, which to-day are confined within the tropics, were then enabled to range to the Arctic circle (Fig. 70).

The great feature of the age was its reptiles. These were represented in all their orders except serpents. Batrachians also existed, if we may judge from some remains found in North Carolina and Pennsylvania in sandstones accumulated probably during this age. These remains belong to the genus *Composaurus*, and reveal, like the Carboniferous Batrachians, some relationship with the Labyrinthodonts. Better characterized Labyrinthodonts have been described under the name of *Centemodon*, from the

Fig. 70. Ideal Landscape of the Age of Reptiles.

same sandstones in Pennsylvania, and one doubtful genus is known in Europe. This was the last appearance of the type. It barely survived till the opening of the Jurassic Age, and then dropped totally from existence.

From the little peninsula of Nova Scotia we have also a Triassico-Jurassic reptile of lizard-like affinities, which Dr. Leidy, of Philadelphia, has named *Bathygnathus borealis.*

The marine Saurians were present in great force. One, *Clepsysaurus Pennsylvanicus*, paddled around the shores of the bays which rested in the valleys of the Alleghanies, while two other genera—*Ichthyosaurus* and *Plesiosaurus*—besides still others of less consequence, made hideous the waters of Central Europe. The animals belonging to the last two genera are among the most wonderful and heteroclitic that ever existed upon the earth. The *Ichthyosaurus* had the general contour of a dolphin, the head of a lizard, the teeth of a crocodile, the sternal arch of an ornithorhynchus, and the paddles of a whale. The *Plesiosaurus* had also the head of a lizard and the teeth of a crocodile, in conjunction with the neck of a swan, the trunk and tail of a quadruped, and the extremities of a whale. This animal was undoubtedly carnivorous, and was adapted for swimming around the shallow margins of coves and bays, and darting its long and flexible neck beneath the surface of the water to seize its aquatic prey. On being pursued by the swift and ponderous *Ichthyosaurus*, it could dive beneath the water and rest upon the bottom, while its serpent-neck reached to the surface, and respiration continued unimpeded (Fig. 71).

But, strange as were the combinations of characters presented by these two animals, they were even surpassed in eccentricity by the *Pterodactyls*, which now first sprang into existence. It was not easy to decide, on their first discovery, whether they bore closest resemblance to mam-

Fig. 71. Ichthyosaurus and Plesiosaurus.

mals, birds, or reptiles; but when comparative anatomy became better understood, it was perceived that their relations to mammals and birds were only in external forms, while the essential features of their structure were undeniably reptilian. Every one has heard of flying dragons, reptiles which, like "flying fishes" and "flying squirrels," are able partially to sustain themselves in the air by means of parachute-like expansions from their bodies. But in the Pterodactyls were true aerial reptiles, as bats are genuine flying mammals (see Fig. 72). The Pterodactyl, in the length of its neck and form of its head, resembled a bird. The trunk and tail were like those of a quadruped. The numerous conical recurved teeth were formed after the Saurian type. The anterior extremities were constructed after the character of bats, the last finger having been greatly elongated, and adapted for supporting a membranous wing, the impression of which is sometimes preserved in connection with the bones. We know twenty species of this remarkable order, all Old-World marvels save a single pair of long finger-bones found at Phœnixville in Pennsylvania. Some were no larger than a snipe, while others were capable of expanding their wings to a breadth of sixteen feet.

Along the valley of the Connecticut River, from the neighborhood of New Haven to the northern part of Massachusetts, is a brownish-red sandstone, resting mostly in horizontal beds, which have been extensively quarried for building purposes. On the banks of the river at Portland, opposite Middletown, are excavations several acres in extent, which have been in progress more than a hundred years. Thousands of ship-loads have been sent down the Connecticut, and built into the aristocratic brown stone fronts of New York. This formation furnished a valuable resource to the earliest settlers of Connecticut. Their

Fig. 72. Haunts of the Pterodactyl.

names and virtues are commemorated on the brown stone slabs still standing in the oldest cemeteries. The sheets of this formation were spread out in an elongated depression in the surface of the older and underlying formations. On each side of this belt of horizontal sandstones we reach a limiting wall of tilted gneisses and bubbling granites. These were the land while the waters of Long Island Sound stretched through an estuary up to New Hampshire, and received there the waters of the embryo Connecticut. If the student of the world's history will go to the Portland quarries, he will see, spread over the ground in the vicinity of some of the offices, slabs large and small, bearing traces like the imprints of the feet of birds. These track-bearing layers of the rock are found at all depths in the quarry. The formation is generally believed to belong to the later Triassic or earliest Jurassic (Fig. 73).

The ornithic character of the footprints has been strenuously argued by Dr. Deane, the discoverer, and Professor Hitchcock, the first describer of these ichnolites. This opinion has been supported by the weight of such names as Buckland, Lyell, Mantell, and Forbes; but all observations hitherto made on the distribution of organic types through geological time tend toward the general principle that every class-type of vertebrates, and every ordinal type of invertebrates, has been introduced upon the earth in the line of succession indicated by its rank, and there is an *à priori* improbability of the existence of so high a type of organization as we find in birds—and birds of the size that these must have been—at a time when the reptilian type had scarcely reached its culmination.

Moreover, the Pterodactyls have made us acquainted with the existence and characters of bipedal reptiles in the very age when the bipedal footprints of the Connecticut sandstone were impressed. It should be noted, also, that the *Rhynchosaurus* of the Trias was a three-toed bipedal reptile, as was also the *Ramphorhynchus* of the Jurassic (Fig. 74); and some three-toed bipedal tracks of the Wealden have also been referred to reptiles. Professor Cope, of Philadelphia, the most accomplished herpetologist of our country, has very recently enunciated the conviction

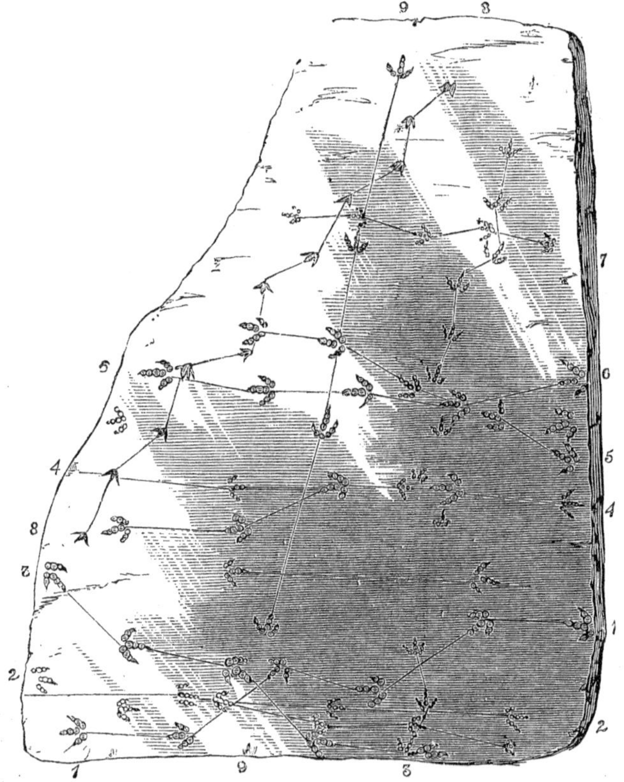

Fig. 73. Ichnolites, or tracks on stone.

[A slab of sandstone (eight feet by six) from Turner's Falls, Massachusetts, impressed with numerous footprints of bipeds, possibly birds. The tracks indicate ten or twelve individuals of various sizes. Discovered by Dr. James Deane, of Greenfield, Massachusetts, 1843.]

that the so-called bird-tracks of the Connecticut sandstone were mostly made by *Bathygnathus*, a reptile to which I have already alluded. One fact, however, of comparatively recent discovery I must not omit to mention. Among the lithographic schists of Solenhofen, in Bavaria, have been exhumed the remains of a vertebrate possessing some of the characteristics of both birds and reptiles. The tail, which is somewhat elongated, after the fashion of the reptiles of the same age, is seven inches in length, and consists of twenty vertebræ, but is furnished with a

Fig. 74. Ramphorhynchus (restored). One quarter natural size.

row of quils along each side. The metacarpal bones are four in number, instead of two or three, as in birds, and the pelvis is also decidedly reptilian. Whether bird or reptile, paleontologists have well hesitated to decide. Professor Dana is fully convinced that we ought to regard it as a "herpetoid" bird, exhibiting a transition from the lower to the upper type, a composite type destined in the next period to be decomposed into two distinct class types. Even if we regard the *Archæoptenyx* as more bird than reptile, and admit that beings of this structure may have lived at one time upon the mud flats of Connecticut, it does not yet follow that the footprints under consideration were impressed by typical birds like those to which these tracks have generally been attributed. [See Appendix, Note VI.]

I am led, therefore, to dissent from the conclusions of Dr. Hitchcock, and to contemplate the tridactyl footprints described by him as the vestiges of reptiles—perhaps ornithoid reptiles—whose exact organization has not yet been ascertained. It is certainly one of the wonders of geology that so many thousands of footprints should have been preserved in the sandstones of Connecticut and Massachusetts, and so very few bones discovered of the creatures which made them. In fact, the only traces of bones thus far known were discovered in 1820 at East Windsor, and publicly noticed by Professor Nathan Smith, and more minutely described in 1855 by Dr. Wyman. These bones were *hollow, like those of birds*, and were thought to yield some support to the bird-track theory. But, besides the presumption that the first birds would not possess this endowment of the higher and typical families of the class, it is well known that many Jurassic reptiles—the *Dinosaurians*—were equally possessed of hollow bones. The imperfect condition of these few remains, however, renders it impossible to decide upon their affinities.

The number and character of these footprints are truly wonderful. Dr. Hitchcock formed a grand museum at Amherst College, containing eight thousand tracks. In his report on the "Ichnology of New England," he figured and described from their footprints no less than one hundred and nineteen species of animals, of which thirty-one are regarded as birds, and forty-seven as reptiles and batrachians. These footprints occur in regular series, extending sometimes a distance of several feet over the exposed surface of the rock (see Fig. 73). Series of tracks of various sizes and species often traverse the same slab. Dr. Deane sent to the British Museum, in 1844, three slabs covered with footmarks, one of which is eight feet long and six feet wide, and contains over seventy tracks made by ten or twelve different individuals. Professor Marsh is at this moment engaged in forming a grand standard collection of these footprints for the museum of Yale College, and has already created a collection second only to that at Amherst.

The largest tracks thus far observed are twenty inches in length, and were made by a reptile which had a stride of three feet, and appears to have walked like a biped, only occasionally bringing his fore feet to the ground. One of the specimens of this species in the Amherst cabinet is a slab thirty feet long, containing eleven tracks. A slab in the British Museum is impressed by footprints fifteen inches in length, forming a consecutive series of five or six, and being from four to five feet apart. Whether bird or Saurian, it must have been a formidable beast to be seen striding along the beach. Such populations once swarmed upon the plains of the Connecticut Valley, now vocal with the hum of civilized life.

It is a solemn and impressive thought that the footprints of these dumb and senseless creatures have been preserved in all their perfection for thousands of ages, while so many of the works of man which date but a century back have been obliterated from the records of time. Kings and conquerors have marched at the head of armies across continents, and piled up aggregates of human suffering and experience to the heavens, and all the physical traces of their march have totally disappeared; but the solitary biped which stalked along the margins of a New England inlet before the human-race was born, pressed footprints in the soft and shifting sand which the rising and sinking of the continent could not wipe out. The blood of the thousands and hundreds of thousands who fell on the hundred fiercely-contested fields of the "Great Rebellion," and the traces of the manful struggles which they waged, were all washed out by the next spring rains, while even the ripple-marks of the age of Saurians, and the impression of the rain-drops of the passing shower, are perpetuated in all their distinctness through ages. Man's history is not written on rocks and river shores. His monuments are not footmarks imprinted on the soil and sands of earth, but achievements of moral and intellectual labor, less perishable than the visible records of the ancient Saurians, because inwrought into the lineaments of the indissoluble soul.

Even the imperishability of the records of the long extinct reptile suggests honor, and encouragement, and hope to the mind of man. For what are these Saurian footprints so carefully preserved, when man is the only intelligence that can duly ponder their significance? Are they not the materials of thought which Providence has kindly stored for a thinking race? words of revelation touching the vast movements in which he has been concerned? gleams of light, which stream far down the avenues of the past, and disclose to our astonished eyes embodied forms moving like spectres of night across the marshes and along the shores of mid-eternity? Well might the heavenly-minded Hitchcock symbolize these teachings by the hinging of a pile of rocky leaves into the similitude of a book. And happily did chance or Providence direct the building of some of the sheets of this rocky volume into the walls of the University at Middletown, where the student, wearied and befogged in the perplexities of human dialects, could look upward to the library-stones of his *alma mater*, and refresh his soul with the interpretation of the language of the Omniscient.

I hasten to the conclusion of my sketch. This reptile-producing age of the world was fruitful in the varied forms of gigantic lizards and crocodiles. To the former belong *Durydorus serridens*, and probably *Sauropus primævus* of the New Red Sandstone of Pennsylvania, and *Bathygnathus borealis* (as before stated) of similar rocks in Nova Scotia. The crocodiles of the earlier epoch of the Jurassic Age came upon the earth in herds. They mostly possessed the peculiarity of having their vertebræ concave before and behind, like those of fishes—a character for which the term *amphicœlian* has been invented by Owen. A few, as the *Streptospondylus*, were exceptional among vertebrates, in having their vertebræ convex before and concave behind (*opisthocœlian*), while the rule among all existing animals of this family is to have the vertebræ concave before and convex behind (*procœlian*).

The most gigantic of all reptiles that ever crawled over the face of the earth or swam in its waters were those of the family of *Dinosaurians*, whose elongated and ponderous forms must grace the picture of Oölitic and Wealden scenes. Of these, the *Megalosaurus* was the advance guard, and measured forty feet in length. The *Iguanodon* and *Pelorosaurus* followed in the Wealden epoch, the former of which was sixty feet in length and the latter seventy! Turtles, the highest order of reptiles, made their advent in small numbers toward the close of the Jurassic Age, but never flourished in abundance till after the reign of gigantic saurians. Just as the curtain was falling on the scenes and actors of this wonderful drama of reptilian life, two or three small mammals ran upon the stage, and gave themselves up to extinction barely in time to enable us to say that the highest class of vertebrates added its contribution to the animal variety of that period in which the Alps were accumulating as sediments in the bottom of the sea. We have honored their memories by bestowing upon them such names as *Thylacotherium*, *Phascolotherium*, and *Dromatherium*, the latter of which was discovered by Professor Emmons in North Carolina, and all of which occupy a low position in their class.

The Cretaceous Age followed the Jurassic, and the Wealden epoch was its first chapter—unless we adopt the late suggestion to annex it to the Jurassic. The herpetology of this epoch has been worked out by that eminent geologist and good man, the late Dr. Mantell. Besides its flying reptiles, and crocodiles, and turtles, here was the jubilee of those enormous saurians just mentioned. The Dinosaurs were characterized by the presence of a medullary cavity in their long bones, as in mammals; by their short-toed feet, like those of the rhinoceros; by their sacrum, composed of five or more vertebræ consolidated, while in all other reptiles it consists of two or less; by the articulating of the lower jaw so as to adapt it for lateral or grinding movements; by the double head of their ribs, and by the elevation of the body from the ground when walking. In all these characters they show an approach toward the class of mammals. The age of mammals was not yet; but it was prophesied and heralded from afar by these few sentences transcribed upon the bulletin of creation. The length of the femur or thigh-bone of the *Iguanodon* was, when full grown, more than four feet and a half, while its circumference around the head was fifty inches, and around the smallest part of the bone twenty-five inches. The teeth were obtusely conical and laterally compressed, so as to present a cutting edge, which was serrated, thus resembling the teeth of the Mexican iguana, from which the fossil reptile was named. It was, undoubtedly, eminently terrestrial in its habits, and subsisted by browsing from the trees of the time, as was the habit of the mastodon of a later period. Twigs of cypresses have been found fossil in its stomach; and Dr. Mantell possessed a jaw in which the teeth had been worn down by trituration of food to half their original length.

With peculiar pleasure I turn now to results of the study of American cretaceous reptiles, which are no less brilliant and no less marvelous than those of Mantell and Owen in the Old World. Thanks to the skill of Dr. Leidy and Professor Cope, both of Philadelphia, the cretaceous beds of New Jersey have been forced to yield up the secrets of their life-history. We now know that while the chalk was accumulating in Europe, the marshes, and jungles, and bayous of the American shores were the scene of as busy and intense a life as swarmed upon the coasts of England, France, or Germany. The Cimoliasaur (*Cimoliasaurus magnus*, Leidy) and Elasmosaur (*Elasmosaurus orientalis*, Cope) presented the form of huge sea-serpents from twenty-five to forty feet in length. The body was swollen out to dimensions exceeding those of an ox, and was furnished with a pair of flippers like the whale. The neck and tail were elongated, and in the latter the tail was flattened, and probably used as an oar in sculling. These were carnivorous monsters, and probably made fierce war upon the feeble representatives of the waning dynasty of fishes. The wrecks of the Mosasaur, of another order of reptiles, are strewn along the ancient coast-line from New Jersey to Alabama, where, at Selma and Cahawba, I have seen fragments of their ponderous skeletons protruding from the face of the limestone cliffs cut down by the Alabama River. The turtles of the period contributed a unique variety to the reptile fauna. Not less than twenty-two species have been described from the cretaceous sands of New Jersey. Nine of these were marine "snapping turtles." One of the latter (*Euclastes platyops*, Cope) had a head twelve inches in length, indicating a "snapping turtle" of the formidable length of six feet. The power of such an animal may be estimated by comparison with the familiar "snapper" of modern times.

But the most abundant of all the Cretaceous reptiles of the Atlantic coast were Crocodilians. At the time of which we speak they must have literally swarmed along what is now the river-front of Philadelphia. They peopled every pool and lagoon along the cretaceous shore of Pennsylvania. The Deinosaurs, however, were the great feature of the bayou and the estuary. Like their kindred of the Old World, they rivaled in bulk the yet future mammoth and mastodon. "They exceeded these," says Professor Cope, "in their *bizarre* and portentous aspects; for some have chiefly squatted, some leaped on the hind limbs like the kangaroo, and some stalked on erect legs like the great birds, with small arms hanging uselessly by their sides, and with bony visage surveying land and water from their great elevation."

One of the most remarkable of these reptiles was the Lælaps (*L. aquilunguis*, Cope), a carnivorous kangaroo-like quadruped twenty-three feet in length. It seemed a rude attempt of Nature to realize the notion of a bird in the framework of an alligator. It walked entirely on its hind limbs, or leaped like the kangaroo. "Its toes were long and slender, and probably similar in number and form to those of a bird of prey. They were armed with flattened hooked claws, which measured from ten inches to a foot in length, and, like those of the eagle, were adapted for grabbing and tearing prey. The teeth were adjuncts in this appropriation of animal life; they were curved, knife-shaped, and crimped or serrate on the margin, and adapted like scissors for cutting" (Fig. 75). This was the most formidable land carnivore of the continent, and second to none of the Old World. It was the *Megalosaurus* of America.

Another of the gigantic reptiles which carried on a war of extermination upon the fields destined to be ensanguined by the battles of Trenton and Brandywine was the Hadrosaur (*Hadrosaurus Foulki*, Leidy). The visitor to the museum of the Academy of Sciences of Philadelphia can not fail to be impressed by the skeleton of one of these monsters mounted in the attitude of browsing from a cycadeous tree. This piece of work is by the eminent restorer of extinct animals, B. Waterhouse Hawkins, Esq., of London, to whose courtesy I am indebted for the photographic view which adorns the opposite page (Fig. 76). The Hadrosaur attained the length of thirty feet. The femur or thigh-bone was sometimes five feet in length, exceeding by more than a foot the maximum length of the femur of the Iguanodon of England, the largest of the hitherto known land reptiles. The fore limbs were less than half the length of the hind limbs. The form of the feet and toes shows that they were poorly adapted for swimming. In its habitual attitude it rested, like the kangaroo, upon its enormous hind limbs and tail. With its supple anterior extremities it reached upward to the foliage of the tree destined to afford it food, and drew the branches down within the reach of the grinding jaws. Not unlikely this land-monster walked at times upon its hind feet, while the ponderous tail dragged behind.

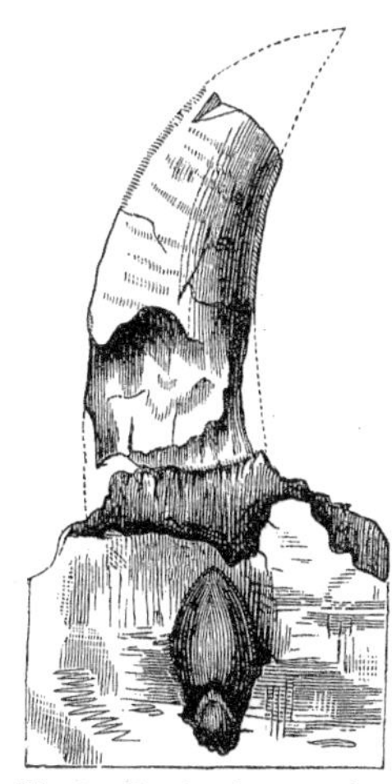

Fig. 75. Tooth of an ancient New Jersey Saurian (*Lœlaps aquilunguis*), showing two successors beneath.

Thus, on both continents, gigantic reptiles long time swayed the sceptre over the entire animal creation. But their empire was approaching its end. One of those throes of Nature, through which new annexations of land to the continent were effected, ushered in the closing stage of Mesozoic Time. The conditions of life were changed; all the peculiar types whose history we have traced dropped out of existence, and mammals assumed the reins of empire. Turtles received accessions of numbers, and serpents now first uncoiled their sinuous lengths, while batrachians made the bayous and marshes resonant with the varied piping of a myriad voices. Among the latter a salamander, known as *Andrias Scheuchzeri*, has attracted most attention, in consequence of having for a long time been regarded as the skeleton of a man, who thus, by the fossilization of his remains, was supposed to attest the reality of the Deluge of Noah.

Fig. 76. The Hadrosaurus.

Such is a hasty glance at the Age of Reptiles. Successive phases of animal life have swept like waves over the surface of our planet, but none has been more striking or more real than that which was dominant through Mesozoic time. Throughout the whole extent of Great Britain there has not been known a single large reptile during the human era; yet in the single era of the Wealden the British dominions maintained four or five species of Dinosaurs fifteen to twenty feet in length, ten or twelve Crocodilians, Lacertians, and Enaliosaurs ten to fifty or sixty feet in length, besides Pterosaurs and Turtles, to say nothing of the probably numerous species whose fossil remains have as yet escaped observation. These successive swells in the stream of animal life are convenient stand-points from which to note the progressive development of organic existence. The history of reptiles, like that of fishes, presents some remarkable exceptional features, which have a most important bearing upon the question of "development," which is taking a front rank among the questions of the age. But these aspects of the case are reserved for future consideration.

CHAPTER XVII.

THE GEOGRAPHY OF THE EMPIRE OF REPTILES.

CONTINENTS have been developed, like organisms, from their primeval germs. Geologic force, like vital force, operates always toward the accomplishment of some definite end; and, notwithstanding its vicissitudes, there is little difficulty in perceiving how every phenomenon of one age has been contemplated and ministered to by the events of all preceding ages. The American Continent is not a single upthrow of volcanic force, but a gradual growth, beginning before the creation of the first animals and plants, and proceeding by a certain method through all the subsequent ages even to the present, and receiving from time to time such progressively improved existences as its physical circumstances permitted. At first it was an angulated ridge of land in the centre of the present continental area (Fig. 20). Then, by successive upheavals, belts of increment were added on the southeast and southwest, till the ancient ocean has been narrowed to its present limits. Like the exogenous growth of an oak, the increase has been always upon the outside. So the vast continent has been built up and configurated in accordance with a method as definite as that which has shaped the globe itself.

The empire of molluscs saw the greater portion of the continent the bed of the sea. The reign of fishes witnessed the emergence of only the extreme northeastern and northwestern portions of the United States (Fig. 55). In the earlier part of the reign of reptiles New England was a peninsula hemmed in by the broad estuary of the St. Lawrence on the north, and a similar estuary of the embryo Connecticut on the west (Fig. 67). Toward the close of this reign the continent had assumed the similitude of its present form and extent (Fig. 77). The Atlantic coast stretched from the neighborhood of New York City to the Delaware River, and thence southwestward to South Carolina, along a line now sixty or seventy miles inland. Delaware and Chesapeake Bays were consequently out at sea, and the Delaware River emptied into the Atlantic at Trenton. From South Carolina the shore-line turned gradually westward, and crossed the States of Georgia, Alabama, and Mississippi at the distance of one or two hundred miles from the present gulf coast. A deep bay set northward along the future valley of the Mississippi River as far as the mouth of the Ohio, or beyond, so that at this time the confluence of those two rivers was at their mouth. West of the Missouri was a vast inland sea or elongated gulf, which stretched along the eastern flanks of the Rocky Mountains to the Arctic Ocean. This gulf was perhaps interrupted at one or two places by spurs of the mountains. Into this gulf emptied the Athabasca, Slave, and Great Bear Lakes. The upper watershed of the present Missouri was beneath the sea; and the basin of the Mississippi was more limited in extent than that of the Ohio, which probably was the larger stream. West of this Mediterranean Gulf was a broad belt of land stretching from the isthmus far to the northwest, and probably to Behring's Straits, if not across them. The Pacific coast was a hundred and fifty miles farther inland than at present. Lake Superior was the only one of the great lakes then in existence. The stream which drained it wound past the future sites of Detroit, Cleveland, and Buffalo, and, plunging over the escarpment near Lewiston, became the ancestor of the present St. Lawrence. The basins of the other lakes are the result of later geological agencies. Probaly large portions of Greenland and other arctic lands had emerged, besides the principal portion of the West India Islands.

The climate of the period was much warmer than that of the same localities in the present age. Coral-builders, and other marine animals now restricted to tropical regions, then flourished throughout the whole length of the continent, from latitude 60° north to the Straits of Magellan on the south. The superior warmth of former ages of the world is probably due, in some measure, to the more highly heated condition of the globe—a source which, through all ages, has been undergoing a gradual diminution. It has also been suggested that the connection which existed between the Gulf of Mexico and the Arctic Ocean permitted the Gulf Stream to flow through the centre of the continent, and thus, while it carried a tropical temperature far toward the north, ameliorated the climate of the regions to the east of it, as the same ocean stream now moderates the cold of high latitudes upon European shores. Thus, while the Northern States were *terra firma*, the rich cotton-fields of Alabama and Texas were gathering their calcareous sediments beneath the Gulf of Mexico. Fleets might have sailed over the rolling prairies of Kansas and Dacotah, and the anchor of the mariner might have fastened in the summit of Pike's Peak. But fleets of *Nautili*, and their cousins the *Ammonites*, were the only keels that plowed that Mediterranean sea, and the polyp and the oyster were the only mariners that cast their anchor on the sunken ridges. Eastward, the broad rolling plains of Illinois and Ohio were adorned with a growth of sub-tropical vegetation, and the west wind of even a winter sky breathed softly over its never-fading foliage. But the shining cities of the West were not there. The kingly alligator alone disturbed the waters of the Ohio. The railroad car, the church spire, the golden wheat, the thronging population—these all were scenes and objects still shut up in the silence and night of the far-distant future. An intelligent being may have stood on the bank of the river, and pictured to himself the shifting scenes of the next half million of years, as we now portray to imagination the expansion of American civilization, and its destined continental grasp of empire a hundred years hence; but no intelligent hand impressed its influence upon the features fashioned by Nature. An occasional voice of monstrous Deinosaur broke the dreadful silence of the broad continent. No song of bird was heard in the grove, and rarely the hum of insect in the air. Bland as the breezes were, and seductive the climate, it was not a fit place for man to be in. Frogs and salamanders must be his pets—lizards and crocodiles his domestic animals. Providence reserved him for a more finished condition of the world.

CHAPTER XVIII.

THE REIGN OF MAMMALS.

ANOTHER cycle of eternity was past. The progress of geological agencies had brought the crust of the earth to a tension which was to be relieved by another collapse. As the Paleozoic Time was closed by the sudden sinking of the beds of the Atlantic and Pacific Oceans, and the corresponding protrusion of the ridges of the Appalachians, so the Mesozoic Time was closed by a farther progress in the same direction. The ever-shrinking nucleus necessitated the ever-enlarging wrinkles of the enveloping crust. The furrows must deepen and the folds must rise. The uplift which marked the close of Mesozoic Time affected the whole continental body. It was not a sudden uprising accomplished in a day. It may have extended through a century; but it was an interval of movements so much accelerated as to mark a pretty definite boundary between two stages of continental development and two great periods in the history of the world. During the Cretaceous Age which had now just closed, the great Mediterranean Gulf represented in Fig. 77 had been broader along its eastern borders, and continuous to the Gulf of Mexico. Through this, perhaps, the Gulf Stream had coursed to the Frozen Ocean. Now, by an upheaval of the central region, this gulf was severed in twain. On the south it retreated to nearly the modern limits of the Mexican Gulf, while northward remained an elongated body of water, swelling out in the central portion of the continent, in two places, to dimensions exceeding the Caspian and

Black Seas of the Old World. Indeed, the area covered by this shallow expanse of deserted and isolated sea-water was the Lectonia of the New World, which, like the level region in the south of Russia, once overflowed by the higher waters of the great seas which stretch along the confines

Fig. 77. Outlines of the North American Continent at the end of Mesozoic Time. The existing boundaries are indicated by dotted lines.
a, a, a. The great Tertiary Sea, stretching from the Arctic Ocean, along the eastern flanks of the Rocky Mountains, to Texas. b, b. The great "Central Plateau," in modern times a *worn-out* continental area.

of the two continents, was destined to be gradually drained. The drainage in both cases was effected partly by the upraising of the continent, and partly by the bursting of barriers and deepening of channels at the point where the imprisoned waters were escaping. But, while the drainage of the European Lectonia was an event which reaches down within the grasp of human tradition, the drainage of the American Lectonia was an event shrouded in the obscurities of the pre-Adamic ages. Thus again we discover that the "New World" is in reality the oldest.

This broad expanse of Tertiary waters stretched across the western part of Dakotah and Nebraska. At the same time, the Mexican Gulf, though outlined somewhat after the modern fashion, was left a hundred miles more extended than at present on the west and north, and reached its long arm up the valley of the Mississippi to the mouth of the Ohio, beckoning to the cooler waters of the North to come and lave its tropic shores. This arm was the southern representative of the Gulf of the St. Lawrence. The peninsula of Florida was a coral reef. A broad belt of the Atlantic States to New York was yet a sea-bottom, and the Pacific yet held possession of the lowland zone of the western slope. Now the Missouri River came into existence, born of the great central sea. Still the Niagara River thundered away in an ancient excavation, which, like the work of the men of Nineveh, was destined to be buried beneath the rubbish of coming ages, and lie a long time unremembered. Now the eastern ranges of the Rocky Mountains were first lifted above the deep. During subsequent ages they underwent further upheavals, while the waters of the Gulf and the great oceans were rolled back to their present positions.

The epoch which followed the last great upheaval, and witnessed the events transpiring on the shores of the American Mediterranean, marked the dawn of the present order of things. All subsequent time has hence been styled Cenozoic. The populations which swarmed upon the earth during each preceding epoch disappeared in turn, and their places were occupied by forms generally more advanced. Of the thousands of species that had their being during the Paleozoic and Mesozoic Ages, not one has survived to the present. The specific types are all extinct. Now, on the contrary, in the dawn of the Cenozoic Ages, a fauna was created, of which a few representatives have survived to modern times. The survivors, however, are all marine. Another feature of the fauna of this era, indicating the approach of the human period, was the advent of multitudes of mammals, a class of which man is the head. Some of the lowest terrestrial mammals seem, it is true, to have made their appearance a long time previously in the Jurassic Age, and perhaps even in the Triassic, but nothing more is seen of the class till the beginning of the Tertiary. Like the Devonian reptiles, they seem to have run far in advance of their class, and to have totally perished for their temerity. The full numerical development and ascendency of mammalian quadrupeds are the characteristics of the Tertiary Age.

The immortal George Cuvier was the first to bring to light abundant relics of these masters of a former world. The vicinity of Paris seems to have been an ancient burying-place of extinct quadrupeds while it was yet the bed of the sea. The bones were undoubtedly transported thither from the adjacent land. One of the most remarkable of these animals was the *Paleotherium*, a three-hoofed quadruped resembling a tapir, and attaining the size of a horse. Other quadrupeds, which grazed upon the same grounds with the Paleothere, were variously allied to the deer, the peccary, and the tapir. Monkeys, mastodons, and elephants existed in Europe a little later, and these were associated with a huge anomalous quadruped named *Dinotherium*, which united characteristics of the elephant, hippopotamus, tapir, and dugong. The sloth and opossum tribes also, which are now confined to other continents, had their representatives in Europe during this period. The prevailing types of quadrupeds were thick-skinned—*Pachyderms*—and cud-chewing—*Ruminants.* The hog and the horse began to exist in the middle of the Tertiary; and somewhat later appear, either in Europe or Asia, the cat, dog, weasel, hare, mink, hyena, camel, antelope, musk-deer, sheep, and ox—of the latter, several species. The *Sivatherium* was an elephantine stag, having four horns and probably a long proboscis. It is supposed to have had the bulk of an elephant, and greater height. This monster dwelt in southeastern Asia. Many other genera, quite distinct from existing forms, have had their former existence disclosed by the patient researches of the comparative anatomist.

America was also a range of gigantic quadrupeds, while the adjacent seas were the abode of mammalian forms allied to the whale. Of these, the one best known is the *Zeuglodon*, whose bones are scattered over portions of the cotton-lands of Mississippi, Alabama, Georgia, and South Carolina. It is a striking sight to stumble over vertebræ a foot and a half long and a foot in diameter, or to see them plowed up from the black soil where they had been mouldering ever since that soil was a sea-bottom. Yet these bones were once so numerous in Southern Alabama that they were gathered and burned for lime, and laid in walls for fences. I have myself seen them used for andirons, and for building the steps of a stile over the door-yard fence. This animal was about seventy feet in length. The skeleton on exhibition in Wood's Museum, at Chicago, is for the most part a genuine representation of the framework of this Tertiary, alligator-like whale. Some of the vertebræ were wanting in this specimen; and in the attempt to restore the missing parts, the paleo-artist has possibly exceeded the bounds of truth, and given us a skeleton of greater length than the facts justify. After a personal and critical examination of the specimen, however, I feel bound to say that this prodigiously elongated creature, that visitors have so long seen coiled about one of the apartments of the museum, is as near a representation of the truth of nature as is likely to be attained. The skeleton possesses one hundred and eighteen vertebræ, of which ninety-one are genuine, and twenty-one factitious. The neck embraces six vertebræ. There are thirty-six pairs of ribs. The cranium is six feet long; the jaws are armed each with five grinding teeth on each side, preceded by two premolars and one incisor on each side of the middle. The epiphyses of the vertebræ—that is, their detached extremities—being unconsolidated with the bodies of the vertebræ, prove that the individual was still immature. This examination was kindly authorized by Col. Wood, the proprietor. We are indebted to Dr. Koch, of St. Louis, for the first restoration of the Zeuglodon, a specimen of which was exhibited, a number of years ago, under the name of *Hydrarchos*, or Water-king, in Barnum's Museum in New York.

Far toward the northwest, on the tributaries of the Upper Missouri, were the cemeteries of American quadrupeds. The shores of the great inland seas already described seem to have been the favorite haunts of the dominant tribes of the continent, while swarms of humbler creatures bathed in their waters, or burrowed in the mud at the bottom. At first these waters possessed all the saltness of the sea of which they were the residuum; but, by degrees, the perpetual drainage, replaced only by fresh waters from the clouds, changed them first to a brackish, and then to a fresh condition. This progressive change is shown by the varying nature of the fossil remains imbedded in the sediments. At the bottom we find the relics of marine animals; in the middle, the vestiges of brackish-water life; and in the last deposits, only relics of fresh waters mingled with washings from the land.

On the White River, in the Territory of Dakotah, in the region where it approaches nearest to the Big Cheyenne, are the *Mauvaises Terres*, or Bad Lands, where Nature seems to have collected together the relics of a geological age, and buried them in one vast sepulchre.

The country to the west and southwest of Fort Pierre, for some hundreds of miles, is an elevated, gently undulating prairie, through which the streams have cut deep gorges for their passage to the larger rivers. It is a vast basin filled with the still horizontal and semi-indurated sediments of an inland sea. The wear of the weather has left many deep scars on the face of the country, and the Bad Lands present us with the mere ruins of a formation which was once continuous. The whole country is treeless and desolate. The soil beneath the feet of the traveler conceals the bones of the numerous populations which enjoyed existence in the earlier Tertiary epochs. The whole scene has the air of the domain of death and solitude. On catching a glimpse of the Bad Lands proper, a most impressive exhibition presents itself. Here, in the surface of a vast plain, is a sunken area thirty miles wide and ninety miles long (Fig. 78). From the bottom of this sunken plain rise domes, and pinnacles, and monuments, and massive walls, which persuade the traveler that he is about to witness the movements and listen to the hum of a vast city. In the language of Dr. Evans—an eminent geologist who almost "dwelt among the tombs" of the ancient world, as they lie stretched out from the Mississippi to the Pacific shores—"these rocky piles, in their endless succession, assume the appearance of massive artificial structures, decked out with all the accessories of buttress and turret, arched

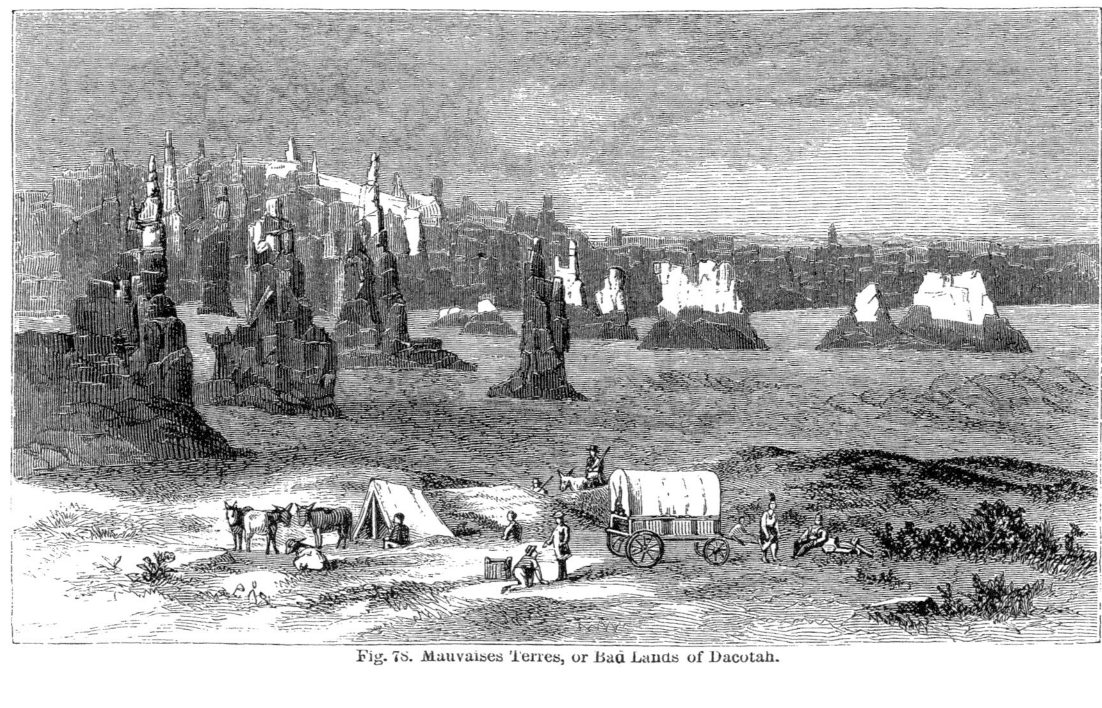

Fig. 78. Mauvaises Terres, or Bad Lands of Dacotah.

doorway and clustered shaft, pinnacle, and finial, and tapering spire."

On a nearer approach the illusion reluctantly vanishes, and all the fancied architecture is resolved into piles of hardened clay and sand. These rise from the bottom of the vale to the height of fifty, one hundred, and two hundred feet, showing along their vertical or sloping sides the varied courses of masonry of which they are composed. In the hundreds of towers and isolated masses that rise from this vale of solitude, the order of the courses is the same; and this agrees with the arrangement in the solid walls which circumscribe the valley. A thousand storms have washed the slopes, and furrowed them into the similitude of fluted shafts and clustered columns, which, at the top, bear sometimes a brown entablature of overhanging grass, or continue upward into tower and minaret. The bottom of the vale is an earth of chalky whiteness, baked by the sun, and utterly destitute of vegetation. The water which oozes out of the foundation-wall of the prairie is brackish and unpalatable. In winter, the wind and snow rush through the lanes and corridors of this city of the dead in eddying whirls, while the withered grasses and the voiceless and motionless solitude, together with the relentless frost and never-tiring storm, make the place the realization of utter bleakness and desolation. In summer the scorching sun literally bakes the clays which have been kneaded by the frosts and thaws of spring, and the daring explorer of the scene finds no tree or shrub to shelter him from the fervid rays poured down from above, and reflected from the white walls which tower around him, and the white floor which almost blisters his feet.

But the most impressive feature of the scene is the multitude of fossil bones which appear built into the massive masonry of this mimic architecture. The wearing and crumbling of the elements roll them out of their long resting-places, and they lie strewn over the bottom of the valley. The traveler feels like one walking upon the floor of a long-deserted and ruined vault. Skulls, and jaws, and teeth, and thigh-bones lie scattered around. Death has indeed held carnival here, and this is the deserted scene of his ghastly repast. But what long ages have glided by since these flesh-covered bones were slain and gathered to the charnel-house! Scarcely a form familiar to the anatomist reveals itself. Here are, indeed, the forms of turtles, large and small, with all the sutures of their protecting carapaces distinctly preserved; but, though turtles, they are unknown species, and some attain a size which, in their present condition, must weigh nearly a ton. Here lie the bones of rhinoceroses—known certainly by their teeth—but different from any existing species. As for the rest of these remains, we do not even know the genera to which they belonged. They present us with strange combinations of characters. One seems intermediate between a tapir and a rhinoceros, while the canine and incisor teeth ally it likewise with the horse. One of the commonest skulls has the grinding teeth of the elk and deer, and the canines of a hog. It evidently belonged to a race which lived both on flesh and vegetables, and yet chewed the cud like our cloven-footed grazers. This has been named *Oreodon*. One of the most wonderful of the beings entombed here is the *Titanotherium*, first discovered by Dr. Prout, of St. Louis. It somewhat resembles a hornless rhinoceros, but is much more massive in its proportions. One of the jaws seen by Dr. Evans had a length of five feet along the crowns of the teeth, and the skeleton of another individual was eighteen and a half feet in length and nine feet in height. Of all the relics uncovered in this ancient cemetery, it is remarkable that but one carnivorous quadruped has been noticed. The fauna of the period was eminently characterized by the presence of pachyderms and ruminants, and this in the same age when Europe was populated by a large admixture of the higher carnivores.

I said this valley of death has the appearance of a subsidence in the wide extended plain. The suggestion is so natural that one almost irresistibly regards it as a vast sunken grave, where the slain of a geological convulsion have been gathered together and decently entombed, and the earth has at last settled down upon their crumbling remains. A better judgment, however, discovers the valley to be the work of excavating waters. Gigantic as the scale of such digging must appear, the geologist is acquainted with other examples immeasurably more sublime. They belong to the phenomena of the Post-Tertiary Age. These towers, then, have not been built up, but have been left in relief, like the figures on the sculptor's marble. Torrents of rain have wielded the instruments that have fashioned the Titanic architecture.

From this Golgotha, if we wend our way northward some hundreds of miles nearer the sources of the Missouri, we find ourselves standing again upon the deposits of a vast inland sea—a sea which was still remaining when the Bad Lands were drained. Around the shores of this far northern basin of water lived, in a later age, the rhinoceros, the elephant, the mastodon, the camel, the horse, the beaver, the wild-cat, the wolf, the land tortoise, and other genera of quadrupeds now extinct. In this lake the Missouri took its rise, while the Yellowstone and other rivers poured into it the drainage of the region beyond, and transported the relics of then existing races, with other sediments, to the burial-place from which they have recently been exhumed.

It gives me great pleasure to make known to the reader

that we are indebted for our knowledge of the details of the geology of these remote and wilderness regions to the energy and science of two young geologists, Messrs. Meek and Hayden, and especially to Mr. F. B. Meek, for the production of paleontological results which vie in thoroughness and exactitude with the best work ever done in any country. In the department of mammalian paleontology Dr. Leidy is our great authority—the Owen of America. These regions were first visited in 1850 by Mr. Thaddeus Culbertson, under the joint auspices of the Smithsonian Institution and his brother Alexander. Later researches were instituted by Professor James Hall, and by Dr. D. D. Owen while in charge of the geological survey of the Northwest, under the auspices of the general government.

Such are some of the phenomena of the Age of Mammals. It was an interval of time when, on all sides of the globe, progressive improvements had brought our earth to a condition suited for higher existences, and the reptiles which reigned in the preceding age were beckoned into the background or driven to extinction. Who that has observed the indications of gradual but systematic advance in animal forms through the ages of the world can resist the conviction that man was contemplated as the termination of the perfecting series?

It is a curious fact that so many genera now extinct from the continent, but living in other quarters of the globe, were once abundant on the plains of North America. Various species of the horse have dwelt here for ages, and the question reasonably arises whether the wild horses of the Pampas may not have been indigenous. Here, too, the camel found a suitable home; but he has disappeared before the intellect dawned which could domesticate him and utilize his instincts. On the Oriental continent the higher types of quadrupeds were now existing, and it looked as if the apex of improvement would first be reached in that quarter of the world.

The uplift of the American and European continents to

Fig. 79. D. D. Owen.

their present levels marked the close of the Tertiary Age. Europe had been an archipelago; but America had long possessed its destined outline, and lacked only the belt which was now added along the two oceans and the Gulf. The continent was now complete. What next could ensue but the creation of man, and the final consummation of the grand work of creation? Human judgment would now have proceeded to the finishing stroke; but Infinite Wisdom saw that the world would be improved by subjecting it to one more ordeal, and then should burst upon it the effulgence of that intellect which characterizes and ennobles the Age of Man.

CHAPTER XIX.

THE REIGN OF ICE.

WHEN the continent of North America, which had been growing through unnumbered ages by continual annexations of land wrested from the dominion of the sea, had finally attained the dimensions and outline destined to endure through the human era—when the great mountain axes had been uplifted, and the broad river streams were rolling the drainage of the valleys and hill-slopes to the sea—when the horse and the camel, the elephant, the bear, and other quadrupeds which were to characterize the epoch of man, had assumed their stations on the land—when the atmosphere was populated by birds and insects which were destined in a coming age to be startled by the presence of a dominant intelligence—when the beech, the tulip-tree, the linden, and the buttonwood had taken their places on the jungle's margin and the highland slope, and the sorrowing willow had begun to weep above the flowing waters of the sedge-bordered stream—when the whole face of Nature seemed fitted and expectant of the crowning work of creation, what should prevent the divine Artificer from summoning man upon the scene to begin the labor of his earthly life? To a finite intelligence the preparation was complete. To the eye of Omniscience one more revolution was needed. The coming man must tarry without the doors of the temple of life through yet another geological æon.

To this time the evolution of the continent had proceeded by elevations and subsidences of the regions lying in the middle latitudes, the resultant of which movements was the establishment of a vast area of dry land extending over all that portion of North America covered by the temperate zone. The northern regions were still the bed of a vast circumpolar ocean. Now, in turn, the high northern latitudes experience an unwonted uplift. Arctic lands raise high their dripping heads above the temperate waters of the polar zone. The climate of the whole northern hemisphere feels the change. No moving currents can now bear torrid warmth to the frozen sea, and return the colder waters to the equatorial zone. The stable land bears sternly the vicissitudes of the clime, smiling coldly in the slanting rays of a summer's sun, and gloaming darkly beneath the auroral shimmering of arctic midnight. The accumulated cold of years binds all the northern latitudes in indissoluble bonds of ice. The northern blast bears frost along the vales which had never felt its power. The limpid streams grow torpid, and then rest in a long hibernal sleep. The verdure of forest and plain, touched by the first breath of winter, shrinks away, and the sere and blackened leaf hangs where there had been perennial green. The ponderous tread of the mastodon turns from the withered meadow to the frozen jungle, and the shivering tapir yields himself a victim to the strange rigors of the climate. The snows of many winters are gathered on the slopes of northern America, and the summer's sun suffices but to change them to a bed of porous ice. Glaciers brood over all the land, and Alpine desolation reigns without a rival over half the continent. Such was the fate of the fair vales which we thought just ready for the occupancy of the human race.

The marks of this stupendous glacier are still visible. As in the glaciers of the Alps, the expansion produced by a summer's warmth would tend to create a motion in the margins of the ice-field. The northern limit was chained by eternal frost to its rocky bed. The southern only was free to move, and the whole expansion would be developed along the southern border. The sliding movement of incalculable tons of ice would plow the soil beneath. Rock-fragments, pebbles, and gravel, frozen in the under surface, were carried forward by the moving mass, while the underlying rocky surfaces were ground away, or polished, or scored in parallel furrows by the irresistible agency of the glacier (Fig. 80). These phenomena are noticeable all over the Northern States wherever the "bed rock" is exposed to view. The bold shore of the north side of Lake Superior has been extensively carved and modified by this resistless action. At Marquette, upon the south shore, are some striking and instructive illustrations. A low dome of metamorphic talcoze schist rises a few feet above the surface of the water at the shore, nearly in front of the Jackson house, which bears the imperishable tracery of its conflict with the continental glacier. The whole surface is smoothed as with a carpenter's plane and sand-paper. The undulations in the surface are scoured as neatly as the level and more prominent portions. Rising from beneath the water on the northern side can be seen numerous grooves and scratches, which glide up the smoothed northern slope, and extend continuously across the summit to the southern side. There are two principal sets of these striæ. One of them extends nearly north and south, the other northeast and southwest. Near this place, and close by the main street as it passes out of town, is an isolated outlying mass of the same kind of rock, which has been left standing out boldly after the destructive agencies that have passed over the surface had plowed away all the surrounding portions of the formation. This stubborn mass stands like a sullen bulwark, defying the most desperate attacks of ice, or

Fig. 80. Great Glacier, Bute Inlet.

storm, or flood. But its lowering brow shows the deep scars of many a fierce conflict. The attacks have evidently proceeded from the north. On this side the perpendicular walls are smoothed and scored in precisely the same manner as the dome-shaped mass to which I have just alluded. The southern side retains, like the other mass, many of the angularities produced by the original fractures of the formation. Similar features are things of every-day observation, but people never suspect what mighty and what extraordinary agencies have been employed in producing them. All our low rocky hills and bluffs are similarly pared off upon their summits and northern exposures, while their southern aspects are more rugged. The great glacier has passed over them, striking them from the north, and grinding down their northerly projections and angularities. These phenomena have been especially studied and illustrated in New England by the lamented Dr. Hitchcock. On the western end of Lake Erie, at Stony Point, the surface of the Corniferous limestone lies two or three feet above the level of the water. Upon this have been deposited four or five feet of gravel and soil. On the immediate shore, the storm-waves have easily washed off the overlying beds, and left acres of the limestone completely exposed to view. What do we find to be the character of this original surface? Level and smooth as a floor—planed down by the energy of the omnipresent glacier—but marked, besides, by some deep furrows, which extend from edge to edge of the uncovered table in lines as straight and strictly parallel as if marked by the "gauge" of some Titanic stone-worker. One set of the furrows, in particular, arrests the attention, since the visitor can not fail to recognize their resemblance to the deep ruts produced by a loaded wagon moving over a soft and clayey surface (Fig. 81).

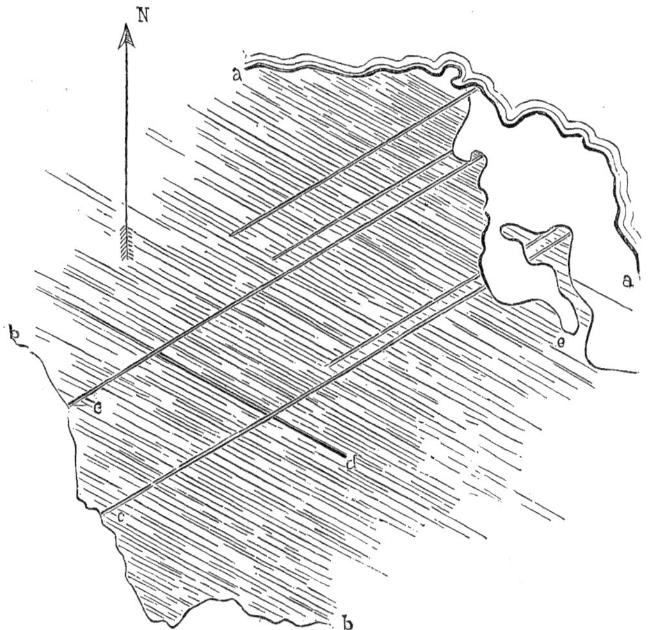

Fig. 81. Sketch of Glacier Furrows and Scratches at Stony Pt., Lake Erie, Mich. a, a. Deep water line. b, b, Border of the bank of earthy materials. c, c. Deep parallel grooves 4½ feet apart and 25 feet long, bearing N. 60° E. d. A set of grooves and scratches bearing N. 60° W. e. A natural bridge.

A result of this wide-spread scouring and grinding of the rocks was the accumulation of vast quantities of detritus. From this source comes a large proportion of the pebbles, sand, and clay which every where underlie the surface-soil, and separate it from the bed-rock—an essential and beneficent provision, as every one knows who has observed the destructive effects of ordinary droughts upon thin soils resting on a rocky basis. Another effect of the great glacier was the destruction of all vegetation over the areas which it invaded. From season to season, and from year to year, the mighty mass marched irresistibly forward, mowing down the forests, crushing tree-trunks, or burying them, with the rubbish of the rocks, from ten to sixty feet beneath the surface. Such buried tree-trunks have thus lain to the present day, and we frequently encounter them in deep excavations for wells, though my friend Professor Lesquereux has strangely asserted the contrary. With other relics of the vegetation of the ancient world were necessarily buried the seeds and fruits of the species then in existence, a fact of which I shall find the use hereafter, in speaking of prairies.

The great glacier moved onward, unheeding equally rocky knob, and swelling hill, and river gorge. I have stated that from the close of the Carboniferous Age the Northern States were dry land. Rains fell, as now, upon the surface, and nourished the vegetation which had found a foothold. The surplus waters gathered themselves, as now, into streamlets large and small, and these, on their way to the sea, wore river-channels in the surface rocks. Across these rivers, across these gorges, the great glacier strode, ignorant of the obstacles to its movement. It bridged Niagara River, it bridged Long Island Sound, and bathed itself in the mild waters of the ocean beyond. It obliterated river-channels, and dug out new ones. It plowed anew the country marked off by the feebler agencies of the preceding epoch. It made a *tabula rasa*, and outlined after a different pattern the topographical and hydrographical features of the Northern States. Many an ancient river-channel has been brought to light by railroad excavations, and more especially by the borings for petroleum that have taken place within the last few years. In many instances the general rocky structure of a region has determined the location of the streams through the same valleys as before the work of the glacier; but even here we find the position slightly varied, and in nearly all cases the present channel is a narrow and shallow one, excavated through the surface of the loose materials which fill the more extensive ancient channel. In Ohio and Indiana these buried river-beds are of frequent occurrence. The ancient gorge of the Niagara River was filled by the obliterating agency of this continental glacier. For ages and ages the river had patiently labored upon this excavation, as it has since done upon the existing one; but the glacier came with its cubic miles of rubbish, and wiped out the trifling furrow, leaving the surface comparatively level, and making it necessary for the river to begin anew its work when the invading glacier had disappeared. The excavation of lake basins is sometimes attributed to this agency, but these may have been partly the result of subsequent aqueous action. It was probably the force which dug the shores of northern seas into their numerous deep and narrow fiords, as can be seen upon the coast of Maine, and the European and Asiatic shores of the Arctic Ocean. It bore southward, over distances of twenty, fifty, and even five hundred miles, fragments of Northern rocks, some of which are of enormous magnitude. One in Bradford, Massachusetts, is thirty feet each way, and weighs not less than four and a half millions of pounds. A boulder of jaspery conglomerate, weighing about seven tons, was transported three quarters of a mile by the class of 1862, and mounted upon the campus of the University of Michigan, an imperishable monument to their memory and their enterprise. The native home of this huge mass is the northern shore of Lake Huron, where the formation is found in place, and where I have seen detached and rounded masses weighing probably a hundred tons. These fragments have thus been transported over lakes, sounds, and seas. Masses of native copper from Lake Superior are strewn over Wisconsin and Lower Michigan, and have wandered even into Ohio and

Indiana; while pebbles of quartz, gneiss, granite, dolerite, and other rocks from the same regions constitute a large proportion of the soil of these states. The streets of Cincinnati are paved with stones which were quarried by the hand of Nature in the region of the upper lakes.

Professor Agassiz, to whom we are indebted for the full exposition and application of the glacial theory, thinks he discovers abundant evidences of the former action of glaciers in Brazil; but the presence of rocky *débris*, and even of rounded pebbles that can not be attributed to shore action, is not enough to establish glacial agency, especially while in the United States we do not recognize it south of the Ohio River. On the contrary, Professor Whitney has recently asserted that the proofs of glacial action are entirely wanting in California, and for some distance northward. The copious accumulations of unsolidified surface materials are attributed to the slow disintegration of the rocks under atmospheric agencies.

Glaciers of almost continental extent still exist on the shores of Greenland, and cover the Antarctic land discovered by the United States Exploring Expedition; also Wrangell Land, very recently discovered by Captain Long in the Arctic Ocean. Perennial ice binds the soil of Northern Siberia, and, as is well known, preserved for many centuries the carcasses of hairy elephants incased in it. There is little difficulty in believing that these high-latitude ice-fields are merely the remnants of glaciers which once extended many degrees farther toward the south.

CHAPTER XX.

LABORS OF THE ICE-BORN TORRENTS, AND THE OCEAN BURIAL.

THE manacles of ice were loosened by the genius of a geological spring-time. Next in the order of vicissitudes was a grand continental subsidence. Vast areas of Northern America, that had been raised to the altitude of perpetual snow, were gradually lowered to the ocean's level. Again the interchange of equatorial and polar temperatures was effected by the moving sea-currents, and the climate of summer smiled over the desolate empire of frost. The rocky glacier yielded to the touch of warmth, and a myriad streams leaped from the bosom of the snow (Fig. 82). Each ice-cold rill united with its fellow, and a deluge of waters set out on their journey to the sea. They wound their way across the future states of Kentucky, Tennessee, and Alabama, to the Gulf. They bore forward a freight of sediments selected from the rubbish bequeathed by the dying glacier, and strewed it over the states that had not been visited by the beneficent action of the ice. Thus the Gulf States and the middle-latitude states shared with the northern regions the materials prepared to serve as the basis of soils in the coming age of thought and industry. These myriad streamlets were, however, unable to bear forward the boulders which had been carried by the ice to the borders of the Southern States. And hence it is that, south of the Ohio, "cobble-stones" are sought in vain. The soil and subsoil possess a degree of fineness and homogeneousness not characteristic of the surface deposits of the Northern States. In the earlier portion of the

Fig. 82. River issuing from a Swiss Glacier.

epoch of thaw and floods, the power of the waters was sufficient to move pebbles of the size of a pigeon's egg. I have observed in Middle and Southern Alabama multitudes of quartzose and other hard pebbles that could not have been derived from any source nearer than the spurs of the Appalachian ridges in the northern portion of the state. One noteworthy locality is along the gorge of the Black Warrior River, at Tuscaloosa, where Sir Charles Lyell, when on his second tour through the United States, mistook them for the "shingle" of the Cretaceous system. This system produces no such pebbles in Alabama. Another locality worthy of mention is at Jackson, on the Tombigbee River, in the southern part of the state, where they constitute a bank a hundred feet in depth. In short, these pebbles may be traced all the way to the Gulf of Mexico; but their normal position is always in the deeper portions of the superficial accumulations. When the power of the transporting currents grew feebler, they bore forward only the finer sands and aluminous sediments which repose generally upon the surface of the Southern States.

The rushing torrents born of the dissolving glacier busied themselves also with the work of excavation. Many an existing valley and river course was determined by the active erosions of this epoch. Many a cut through the rocky ribs of mountains had now to be executed to make way for the escape of imprisoned waters. Many a broad and rock-floored valley became filled, and converted into an alluvial plain, by the rubbish which the torrent deposited in its quieter mood. Many a basin was now scooped out which, in the next epoch, became a lake of standing water. The basins of all the larger lakes that have been excavated by erosive action conform in their longitudinal extent to the strike of the underlying formations. A line running through the centres of the great lakes from Chicago to Oswego, runs approximately along the winding strike of the formations of a certain age. This line shows the configuration of the shore of the continent when those formations were accumulating. It is worthy of particular note that the shore-line was always substantially parallel to the axis of these fresh waters during all paleozoic time. In the Lower Silurian it lay to the north of these waters. During the Devonian it was to the south of the waters. During the Upper Silurian it was to the south in the eastern region, and to the north (or northwest) in the western region. We may here seize upon a *key* to carry with us, and unlock at any time the geological map of the country before the mind's eye (compare Fig. 58). Every one locates instantly and definitely the Niagara Falls and Niagara River. The Niagara limestone was named from the falls, and its outcropping belt trends east and west at that point. This is the great limestone mass of the Upper Silurian. As in

New York, the growth of the continent was toward the south, the rocks of the Lower Silurian must lie to the north of Niagara Falls, and the rocks of the Devonian to the south. From either of these regions trace a line parallel with the axis of the lake waters—omitting Lake Superior—and we have the geographical boundary of a system of rocks, or one of the shore-lines of the ancient continent.

It is a curious fact that the great lakes were excavated along the outcrops of the formations instead of across them. It is not an unaccountable fact, for the lines of least resistance must have run along the trends of the most friable strata. Lake Michigan is scooped out from Devonian formations; and the same is true of Lake Erie. Lake Ontario is excavated in Lower Silurian strata; and the same is true of Georgian Bay, Green Bay, and the Wisconsin lakes farther south—Winnebago, Horicon, and Albion. The basin of Lake Huron is underlaid by Devonian and Upper Silurian rocks. It seems to be two basins coalesced; and but for the peninsula of Niagara limestone separating it from Georgian Bay, it would be three basins blended in one. Lake Champlain also conforms to the trend of Lower Silurian strata, but the small meridional lakes of Central and Western New York are plowed across the formations. They are a kind of inland fiords, worked out perhaps rather by the action of the glacier than by that of the floods which followed.

The influence of these vast inland accumulations of fresh water upon the comfort and happiness of man is strikingly beneficent and providential. They serve as equalizers of summer and winter temperatures. In winter they may be regarded as vast reservoirs of warmth—great natural stoves or heaters, which continue to impart their warmth to the frigid winds that move over them, and thus transfer their influence to the contiguous lands. This is a provision which, till very recently, has been overlooked. It has been well understood that the Atlantic ameliorates the climate of Western Europe, and the Pacific that of Western America. I have had occasion to ascertain that a similar influence is exerted by the great lakes, and to an extent which is far more than proportional to their volume, as compared with one of the oceans. I have investigated the climate and productions of the belt along the eastern side of Lake Michigan, from St. Joseph to Mackinac, and especially in the "Grand Traverse Region," where the bays penetrate far inland, and thus augment the climatic influence of the water. In the Grand Traverse region the thermometer never sinks more than fourteen degrees below zero, and hence none of the more delicate fruit-trees ever suffer injury from the severity of the winter. Autumnal frosts are delayed till late in October, and hence the season is sufficiently long for the ripening of peaches and grapes. Snow falls in November or December, before severe freezing weather arrives, and hence the ground is never frozen, and tender roots stand out through the winter. In extreme winter weather the eastern shore of the lake is from fifteen to twenty degrees warmer than the immediate western shore. But the western shore, as that industrious physicist and archæologist, Dr. I. A. Lapham, has shown, is sensibly milder than the interior of Wisconsin, so that the ameliorating influence of the lake upon the climate of Michigan becomes strikingly manifest. No Northern state can compete with Michigan in the production of fruits. This fact, to a great extent, is owing to its environment by the great lakes. The western slope of the state is most favorably circumstanced in this respect.

Lake Michigan is a body of water three hundred miles long, sixty miles wide, and eight hundred feet deep. The bottom is warmed by the internal fires of the earth. The water stands at least fifteen degrees above the mean temperature of the year in the same latitude. But, even without this warming influence, the mean of the climate is considerably above the freezing point, and the cold of winter does not suffice to depress so large a body of water to thirty-two degrees. The lake, therefore, never sinks below thirty-eight or forty degrees. The bitter westerly winds, consequently, in sweeping across the lake, experience a material softening before they strike the Michigan side. It is worthy of note that, throughout the Northwest, the severest winter winds come from the west and southwest. It is for this reason that the eastern shores of the great lakes are more benefited than the western. As the bitterest winds of all are from the southwest, it follows that a situation which, like the Grand Traverse region, can receive the winds that have traveled the longest distance over the lake, will be best protected from the frosts of winter.

It is probable that the Canadian region, along the eastern shore of Lake Huron, enjoys a winter climate similarly exempt from destructive extremes. The influence of these lakes is sensibly felt even along their northern shores. The region south of Lake Ontario has long been celebrated for its fruits, while the southern shore of Lake Erie has been proven one of the best grape-producing districts of the world.

Such, then, are some of the beneficent results of an incident of the epoch of the dissolution of the glacier. The ice was rapidly melted; torrents sprang into existence, and scooped out lake basins; these became filled with waters which, besides subserving the interests of navigation, exert, perhaps, a more beneficial influence in ameliorating the condition of man in the centre of the continent.

A different ordeal still awaited the destined dwelling-place of American freedom. The subsidence which had restored the genial climate of the Tertiary Age extended from the Arctic to the Temperate regions. By degrees, Wisconsin, Michigan, New York, Ohio, and other Northern States disappeared beneath the waves. Here the denizens of the land had held undisputed sway during the long ages since the coal-bogs were made vocal with the croaking of gigantic Batrachians. Now all that had been gained was lost. The trident of Neptune waved again where stately trees had reared their palmy heads, and the mastodon had hurried through the forest with his thunder-waking tread. Such are the fortunes of contests in the natural as in the social world.

CHAPTER XXI.

RESURRECTION OF THE CONTINENT.

IT seemed like a failure of the plan of creation. The land gained by unnumbered throes of the continent was lost. The higher summits only held their heads above the level of the careering waves. Deposits bearing the marks of oceanic action reach to an elevation of six thousand feet on Mount Washington, two thousand or more on the Green Mountains, and three thousand on Monadnock. But this deep submergence was not of long continuance. Slowly the continent rose again from its deep sea-burial. As summit after summit lifted its gravel-covered brow above the sea, the retiring waves, lingering, dallied with the pebbles on the widening beach. As the continent rose, every inch became, in succession, the ocean shore, and was subjected to the assorting action of the waters. As a consequence, the finer materials were left upon the surface, and a most suitable substratum for the soil was thus provided. During the preceding epoch, Nature robbed the Northern States of their finer material for the benefit of the Southern. Now she made amends by raking up the deep deposits, and selecting and strewing over the surface a new supply of finer detritus for the benefit of the Northern States.

Before the resurrection was completed, Nature made several pauses in her work, and the sea was permitted to stand, perhaps for ages, over districts that had been marked out as the dwelling-place for man. The first pause occurred when the waters still stood four hundred feet above their present level at Montreal. At lower levels, down to twenty-five or thirty feet, the traces of standing waters have been observed about New England and Long Island. At one time the Atlantic flowed up the valley of the St. Lawrence to Montreal, and whales sported in an arm of the sea which reached over the valley of Lake Champlain. The ancient beaches have been traced all around those earlier borders of the land.

The last portion of this upward movement has been in times comparatively recent. We are neither to suppose that the work was suddenly and violently performed, nor that it is even yet complete. The secular elevations now known to be progressing at various points along our coast are but a continuation of the action which rescued our continent from the jaws of the ocean, and which may be farther continued for many centuries. Who knows how much land may yet be added to the northeastern border of America? Who can say that Newfoundland may not yet become a peninsula joined to the main land, or that the ancient submerged prolongation of our continent may not be again resurrected? New England may cease to be "little New England," and may boast of as many acres as the "Great West"—or at least that portion of it covered by the organized states. However, New Englanders ought not to indulge too sanguine expectations in this respect.

Around the Gulf-border of our country the indications of future extension are of a more reliable character. In one region the delta of the Mississippi is continuing to push itself seaward. Materials are being transferred from the Rocky Mountains to Louisiana. The Mississippi is annually building out into the Gulf. From the same source arises another and an unexpected development of land upon another portion of the Gulf-border. Vast quantities of the finer sediments of the Mississippi are floated out into the range of the Gulf Stream, and borne onward around the peninsula of Florida into the Atlantic. As is well known, the Gulf Stream in this region pursues a course from west to east, and, in passing the Florida Keys, it bends northward. At this flexure of the stream, the outer portion of the current must necessarily be more rapid than the inner, or that nearest the main land. The retardation of the inner belt gives more time for the deposition of its sediments, and we accordingly find a submarine bank of mud gradually raised. When the summit of this bank reaches to within about two hundred feet of the surface, the coral-builder plants his foundations upon it, and patiently rears his massive reef to the ocean's surface,

> "Unconscious, not unworthy instrument
> By which a hand invisible is rearing
> A new creation in the secret deep."

In the mean time the ocean stream is crowded farther toward the south, where the waters are deeper. Simultaneously the mud deposits extend southward, and upon these stretches southward also the "masonry imperishable" of the little coral architect. As with all coral reefs, those of Florida are gradually being covered with the materials of a soil, and clothed with a tropical vegetation. Thus the land at this point is continually extending itself. When we inquire for what length of time and over what distance this growth of the land has taken place, we find that half the peninsula of Florida is underlaid by a reef which is absolutely continuous with that now forming, and that the species of polyps which worked upon that foundation of a state were identical with those now laboring to extend the area of American freedom. It appears, therefore, that the same processes which have resulted in the formation of that peninsula are still extending it southward. The time can not be infinitely remote when the "ever blessed isle" will be peacefully annexed to the dominions of the American eagle.

These are events and phenomena whose history reaches down to the present, and whose promises extend into the future. Let us turn back our thoughts for a moment, and reinspect the phenomena and results wrought out by the ocean on occasion of his last supremacy over the land. It seemed, indeed, as if the work of Nature had proved a failure; but this very inundation had been embraced in the plans of infinite Beneficence. I have already alluded to the assorting action exerted upon the loose materials left upon the surface by the retiring glacier. Large portions of the drift were completely worked over, and redeposited under a semi-stratified arrangement. Who has not stood in a railroad-cut through a bank of these materials, and witnessed the bands of variously-colored sands and clays exposed in the walls of the cut? From such an exposure of the internal structure of these hills and ridges one may learn that they consist of beds of clay of various extent, and variously inclined in reference to each other, between which the spaces are filled up with sand and pebbles. Now this circumstance, accidental as it seems to be, has contributed immensely to human convenience. The rains which fall upon the surface of the earth percolate through the superficial layers of sand and gravel, but always, sooner or later, reach one of these strata of clay, by which the farther downward progress of the water is arrested. Upon the top of such a bed of clay the water accumulates. It saturates the overlying sands. It is true that it will slowly follow the descent of the clay bed, and will reach its margin, and begin another descent. It will soon be arrested again in a similar manner, and will form a deeper-seated reservoir, which in turn will overflow and contribute to a third. Thus every clayey stratum, whether of great or small extent, is a natural cistern, where Providence saves the rains within reach of the surface—a cistern of filtered water, preserved in a cool and protected situation. Man penetrates the drift a few feet at any place, and opens one of these natural cisterns and supplies his wants (Fig. 83).

But the dumb beasts have never learned to dig wells. Observe that Providence has not neglected them. The geological forces that have dug river gorges, and scooped out valleys large and small, have cut across these beds of clay, and tapped a myriad cisterns where their contents

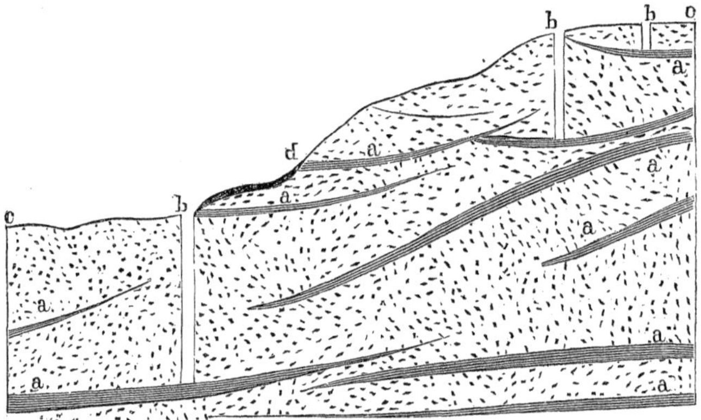

Fig. 83. Phenomena of Wells and Springs in Drift Materials.
a, a, a, etc. Beds of clay variously disposed in a mass of sandy materials. b, b, b. Wells sunk in different situations, and finding a supply of water only when a bed of clay is reached. A well on the top of a hill may be shallower than one at the foot. c, c. The surface of the earth. d. Outcrop of bed of clay, causing a spring. If the porous materials contain fragments of limestones, these spring waters are hard, and deposit travertine from d toward b. A well carried below its supplying-bed may lose its water again.

escape upon a hill-side (Fig. 83), and form a spring at which the untutored brute may slake his thirst without the benefits of shovel and pick. But as all animals could not conveniently resort to springs, and as there are certain regions that have not been scored by denuding forces, we find the hill-side spring wandering off in a modest rill. At length it joins hands with a neighboring rill, and, with augmented force, they urge their way downward. Soon they are joined by other streamlets, and thus the united waters form a rivulet, which goes wandering about the country, seeking the thirsty and weary brute, and dispensing its blessings to all that choose to partake. On this traffic no excise-tax has been imposed. The rivers are still free to perambulate the country, and furnish entertainment and comfort for man and beast.

Suppose, for a moment, the surface of the drift had been left an unbroken plain. The native cisterns might still have been inclosed, but no leakages could occur; neither spring nor rivulet could originate under such circumstances. A stream originating in another region might flow through this, but even such a stream would diminish instead of augmenting its volume. The greedy sands would consume it. Like the rivers of Nevada, its beginning would be more imposing than its end.

But suppose, farther, that in a country of such unbroken surface, the argillaceous particles had not been separated in beds of clay. Suppose the entire mass of drift materials a promiscuous mixture of coarser and finer constituents. What would become of the water precipitated from the clouds? It must descend to the rocky foundations of the land. Man, who would seek a well, must dig to the bottom of the formation. This might be one, two, or five hundred feet. Such a condition of things would be rather inconvenient. And yet the existing condition results from an incident in geological history which, at first view, seems only destructive and retrogressive.

Nor is this all. Vegetation has greater need than man that a large body of water should be held within a limited distance of the surface. Were rains always frequent, this necessity might not exist; but large portions of the earth's surface are subject to droughts of greater or less severity.

Now in time of drought, when the immediate surface has become parched, moisture rises by capillary attraction from the nearest subterranean reservoir. Evidently the nearer this reservoir to the surface, the more easily and rapidly its alleviating effects will reach the suffering roots of vegetation. So far as I can perceive, the soils of regions circumstanced as I have supposed must be doomed to irremediable sterility.

The epoch of the last emergence of the land was the time when the precise drainage features of modern times were determined. The great undulations of the surface which determine the principal water-sheds depend, it is true, upon the conformation of the rocky crust, and probably existed nearly as they now exist in the age anterior to the reign of ice. But all the subordinate details of the drainage were executed while the continent was rising from its last ablution. There was a time when the descent of the Mississippi was so gradual that the waters spread out many miles, and, like a vast bayou, filled the valley between the bluffs which bound to-day its alluvial bottom. This is the condition of the Amazon in the existing epoch. As the interior of the continent became more elevated, the currents of the rivers became swifter and narrower. Many river-channels, obliterated by the agency of the glacier, and, farther, by the blending and leveling of the submerging waves, could not again be found when the continent was restored. The new-formed streams were obliged to seek out new paths, and dig new outlets to the sea. The ancient gorge of the Niagara had been obliterated, and, when the labors of that stream were resumed, a slight change in the configuration of the surface turned the current from "the whirlpool" farther to the east than before. On reaching, near Lewiston, the brow of the escarpment which then formed the head of the great Laurentian estuary, it missed its ancient channel, and, plunging headlong down the precipice, began again the practice of that stupendous system of engineering which it had already so well learned to wield.

As the continent slowly rose from the sea, innumerable depressions in the newly-exposed surface were left filled with the brine. Thus the basins of the great lakes of North America were first filled. But an outlet existed from these lakes to the ocean. When the accession of water from the clouds produced an overflow, the drainage was always saliferous. Thus these lakes have always been giving out brine and receiving only pure water. As a consequence, their original brine has been continually diluted, until, in our age, its salinity is no longer perceptible to the taste. Nevertheless, chemistry has a tongue that still detects the salty savor.

Not a few of the ancient depressions found no outlet. The ocean's brine, imprisoned within impassable barriers, has there remained, and "salt lakes" are the result. In many instances the brine of these lakes has even been concentrated in the progress of time. The evaporation of pure water from their surfaces has exceeded the precipitation from the clouds within the limits of their hydrographical basins. This is probably the case with most existing salt lakes, of which the Caspian Sea is our largest example. Some of these salt lakes, in the progress of evaporation, have greatly shrunken in geographical extent. Their abandoned territory is often saturated with saline constituents rejected by the overburdened water. Some of the salt and alkali plains of our Western Territories have had an origin of this kind. It can hardly be doubted, however, that the great salt-plantations of Nevada result from dried-up streams which take their origin in salt-bearing formations built into the frame-work of the sierras.

We have now arrived at a point where we can read the

history of our beds of peat and marl. These neglected swamps demand a better appreciation. Improved machinery is already offering us peat for heat-production. There was a time when the richest coal-bed was but a bottomless peat-bog. The coal-measures of the country are nothing but fossilized "swamp-lands." Nature has shown an interest in peat. Let us see how she prepares it in modern times.

I have already called to mind the grand events which accompanied the last great revolution of the globe. We have seen, in imagination, the world emerging in a resurrection from its grave of waters. The waves have glided down the shoulders and sides of the continent until she sat with her feet only bathing in the sea. But the surface of the land was covered with inequalities, and thousands of little depressions held their lakelets of water prisoners in their arms. So the land was at first dotted with thousands of little inland seas. How some of them, with no outlet, held fast to the saltness which was the last bequest of their mother ocean, I have already explained. How others, like spendthrifts, permitted a perpetual outgo, with no income to correspond, I have also reminded the reader. At what particular stage of dilution Nature ceased to regard them as fitting abodes of the marine animals which must have been entrapped within their borders I am unable to say. By what means they became tenanted by the beings which make their home in fresh waters I am unable to say from the observed operation of natural laws. I have no doubt that Nature promptly produced, *ab origine*, such creatures as would be suited to the new circumstances.

But the history of multitudes of the smaller and shallower lakes has been completely closed. For ages they received and swallowed up the leachings of the surrounding hills, and their generally calcareous waters precipitated, by degrees, a bed of fine calcareous mud. To this were added the dead shells of myriads of little molluscs that flourished upon the lime held by the waters. The bottom of each lakelet became a bed of marl. But all around the margins of the lakelet the grasses and sedges were vying with each other in venturing into the water. The amphibious rushes put them both to shame by raising their dirty heads sheer through the slime of the lakelet's bottom. And there they stood—the rushes up to their knees in water, and the sedges and grasses scarcely over shoe. And every leaf and stem which fell upon the water or found its way to the shore, became entangled in the herbage, and lay down and rotted there; and the rush, and the sedge, and grass, when shrill November came,

> "With wailing winds, and naked woods, and meadows brown and sear,"

bowed their heads in his presence, and wrapped themselves in the cerements that had gathered about them. Thus a soft bed of vegetable mould fringed the lakelet, and overlapped the deposit of marl which was growing beneath the water. From year to year, as the water shallowed about the margins, encroaching vegetation crowded farther and farther toward the centre of the lakelet. I have not seen the beginning of this process; but at that period of time in which I have been permitted to begin my observations, I find these changes in progress. I have detected Nature *in mediis rebus.* The little herb standing by the water's brink this year, dies, and forms a deposit exactly like that which was formed in the year before my eyes—or any human eyes—detected the character of these vicissitudes; and my logic compels me to reason from that which I have seen to that which no man has seen. And so it is of the changes upon the ocean's shore, until the facts of the passing world are made to illuminate the dark and mysterious chambers of the fossil realm.

Reasoning thus, we are forced to the conviction that many of the ancient lakelets have become completely filled. Others are only half filled. Others have had the work completed even "within the memory of the oldest inhabitant." Who is not acquainted with some grassy pond which his father had known as a clear lakelet? What man is unable to point out some swale that in boyhood he had known as a grassy pond? or some meadow that he has traversed as an old-time swale? The work is not ended when the lakelet is filled. The surrounding eminences still continue to afford lime-yielding water, which saturates the muck and deposits its lime; while vegetation still pays its annual tribute to the accumulating stores, till the solid material becomes sufficient to exclude the excess of water. The ancient lakelet is at length a finished meadow. Man now steps in and appropriates the annual crop as coolly and unthinkingly, and perhaps as thanklessly, as if kind Nature had not expended a thousand years and infinite pains in fitting it up for his uses.

The epoch of the resurgence of the continent has been styled the Champlain Epoch of the Post-Tertiary Age. During this epoch existed the mastodon and mammoth, whose ponderous bones and teeth have overstrewn the entire area of our country. Unlike the teeth sown by Cadmus, those of these giant quadrupeds produced no crop, and we are not early enough in our visit to this planet to be gratified by the exhibition of living mastodons and hairy elephants.

It was probably also in the earliest part of the Champlain Epoch, or even before the full termination of the Glacial Epoch, that man appeared upon the earth. Judging solely from geological data, his appearance in America was considerably later than in the Old World; but even in America the race has probably looked upon the later representatives of the mammoth and mastodon tribes. I have myself exhumed mastodon bones from a bed of peat not more than three feet deep, and which I believe could easily have been accumulated during the last five hundred years. The traditions of the American Indians in reference to the acquaintance of their ancestors with animals which left these gigantic remains are probably founded upon fact.

But this is a subject to which I shall return. I am tempted still to offer a few farther reflections upon the physical events marking the dawn of the Human Epoch.

CHAPTER XXII.

FORMER HIGHER LEVEL OF THE GREAT LAKES.

IN the spring of 1865, at the time of the memorable floods, I had occasion to pass over the Great Western Railway from Suspension Bridge to Detroit. From Chatham to the vicinity of Detroit this road runs within sight of Lake St. Clair. On this occasion the country was submerged almost as far as the eye could reach in every direction. Our engineer seemed to be practicing a new species of navigation—rather grallatorial than natatorial. The little lake had become rampant. Outraged by the long encroachments of the land, it had decided to assert again its ancient supremacy. Then I was reminded, if I had never been before, how slight a rise in the lake would submerge entire counties lying upon its borders.

A large part of this Canadian peninsula is scarcely above the ordinary level of the lakes. The whole region looks like an ancient swale and a more ancient lake bottom. The same is true of a considerable breadth on both sides of Lake St. Clair and the Detroit and St. Clair Rivers. Lake St. Clair itself—except when rampant—is little better than a marsh with a river running through it. Among navigators it is the opprobrium of the lakes. One never ceases to hear sailors talk about "the flats," and Congress never ceases to be importuned to make another lake where Nature is in the very act of blotting one out. If the reader has ever taken a steam-boat trip through the lake, he could not avoid discovering that it is the very similitude of ostentatious learning—"all breadth and no depth." The bullrushes are boldly invading and occupying it on every hand. A thousand incipient islands are breaking up its continuity. Once it was fifty miles in width and a hundred miles long. A rise of ten or twenty feet would make it that again.

But the whole series of lakes is nearly of the same level from Chicago to Buffalo. The former high waters of Lake St. Clair imply similar floods in the other lakes. Indeed, we easily discover corroboration of this in the topography of the country at Chicago, Detroit, and Toledo. These cities are built upon the slime of the lakes, and a slight elevation of the waters would bury them beneath a new deposit of lacustrine mud. The artesian wells of Toledo are supplied from some of the sandy beds of the ancient lake sediment, which follow the general configuration of the underlying drift, and come to the surface at some higher level back of the city.

These evidences of higher waters lead us to inquire for the cause. They could scarcely be occasioned by a greater volume of water, since the outlets are of sufficient capacity to prevent its accumulation. Nothing but an obstruction of the outlet can explain the phenomenon. This obstruction must have existed at a point where the contiguous shores were sufficiently elevated to prevent a flank movement of the water. It must also have existed at a point beyond or to the eastward of all these obvious traces of the inundation. It could not have been at Mackinac, for that would not have flooded Canada West. It could not have been at the foot of Lake Huron for the same reason, and because the contiguous country is too low. It could not have been at Buffalo for the last-named reason, and also because the country between Buffalo and Lake Ontario belongs to the submerged area. It must have been at the mouth of the Niagara River.

I have said the Niagara River commenced its present gorge during the Champlain Epoch. In reality there was no Niagara River when this work commenced. Lake Erie stretched down the valley of the existing river, and the overflow of its basin wore the notch in the rocky rim which was the beginning of the Niagara River.

Lake Erie stands at present three hundred and thirty-four feet above Lake Ontario. At the time of which I am speaking it stood three hundred and seventy-two feet above Lake Ontario, and filled the valley of Niagara River as far as the heights above Lewiston (Fig. 84).* Indeed, there are clear evidences, in the form of beaches containing fresh-water shells, that the level of the river was once forty feet above the present summit of the falls. No barrier has ever existed to dam the water to this height except the escarpment at Lewiston. This is one hundred and five feet above the summit of the falls, and thirty-eight feet above Lake Erie. The indications seem to be conclusive that the waters of Lake Erie stood thirty-eight feet higher than at present, and poured over the bluff at Lewiston, in a series of cascades, three hundred and seventy-two feet, to the sea, which at this time filled the basin of Lake Ontario. During the subsequent ages, the mighty stream has dug a gorge in the solid rock, which is seven miles long, two hundred and fifty feet deep, and, on an average, about one thousand feet wide. The material transported from this gorge into Lake Ontario is over three hundred and forty millions

* *Explanation of Fig.* 84.—The diagram on the following page is intended to illustrate the geological position of Niagara River and Falls, and the ancient lake levels from Lake Ontario to Chicago. The vertical scale is 560 feet to the inch; the horizontal scale is irregular. The diagram is merely a series of sections around the lakes, placed end to end. The dips of the strata are much exaggerated. The two portions of the diagram join each other along the line *a*, *b*, *c*, *d*, etc. The figures against the vertical dotted lines show the heights in feet above the sea of the points to which the lines extend.

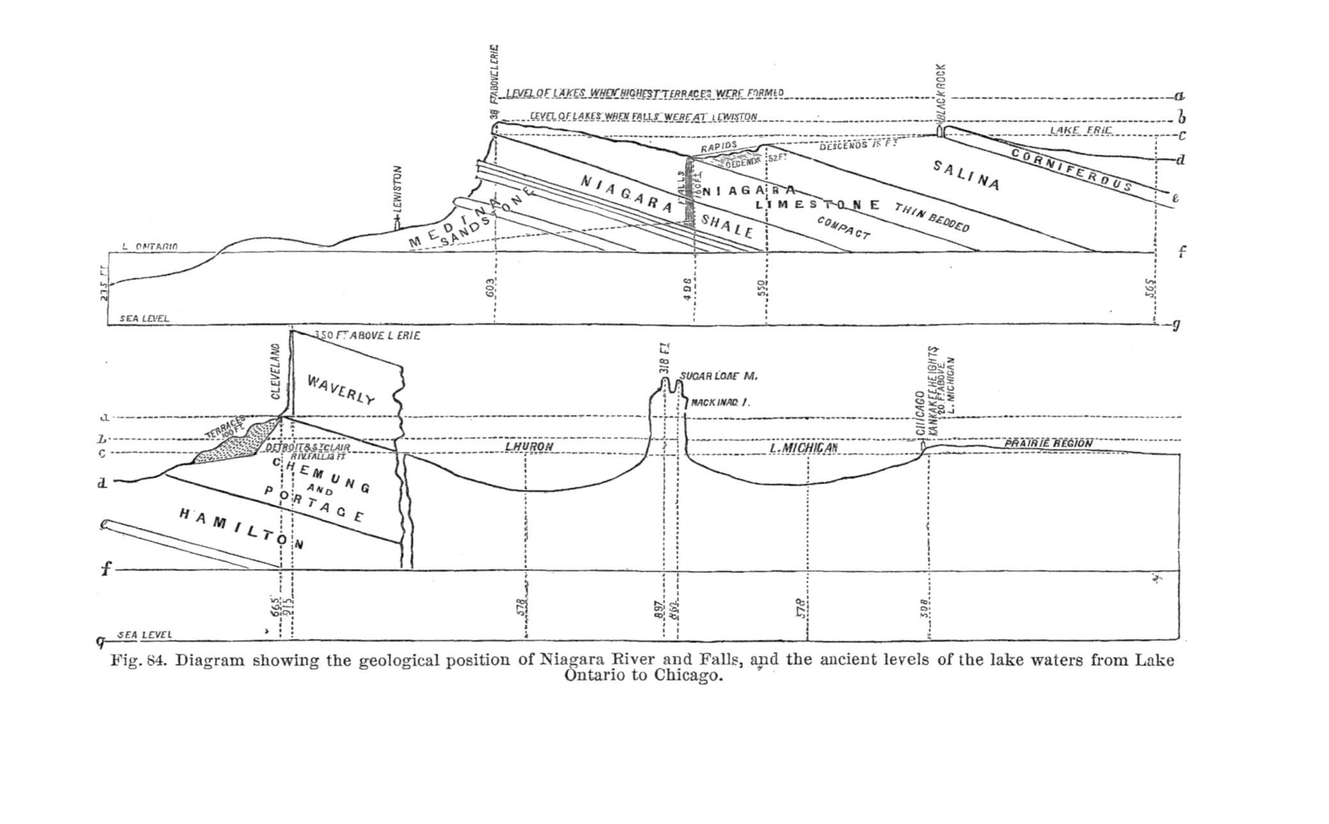

Fig. 84. Diagram showing the geological position of Niagara River and Falls, and the ancient levels of the lake waters from Lake Ontario to Chicago.

of cubic yards, and weighed nearly seven billions of tons. The time consumed in the execution of this stupendous piece of engineering may be roughly calculated from the observed rate of recession of the falls. In 1842 Professor Hall executed a careful trigonometrical survey of the shore-lines and landmarks of the falls. In 1855, twenty-three years later, M. Marcou made careful re-examinations, which he reported to the Geological Society of France. From these data it appears that the Canadian Fall, over which the largest body of water is discharged, has receded, by the wearing of the rocks, to the extent of twelve feet, or a little more than six inches a year. With this clew, we determine that the time required for the excavation of the entire distance from Lewiston is over seventy thousand years. This presumes the rate of recession has always been the same. The more I consider this subject the more I am impressed with a conviction that the rate of recession was formerly more rapid than during the last one hundred years. I am willing to reduce the time consumed to twenty thousand, or even to ten thousand years. Geologists most greedy of time ought to be satisfied with this when it is considered that this interval is but the unit in the arithmetic which calculates the time consumed in the revolutions of the globe. Before the beginning of the excavation of the great gorge, geological agencies had strewed the surface with drift-deposits, some of which had been transported hundreds of miles. Before the transportation of the drift, the basin of Lake Ontario had been scooped out, and the vast erosion of the escarpment at Lewiston had been effected. Before the period of the erosion was that of the solidification of the sediments; and still farther back, the incalculable intervals during which the sediments were accumulating five miles of thickness. At the commencement of the excavation of the gorge, the fauna which populated the region was essentially the same as now. If, in an interval of twenty or even of ten thousand years, little perceptible change has taken place in the populations of the globe, how vast a period must have elapsed during the progress of organic mutations which have twenty times resulted in the almost complete extinction of existing forms, and their replacement by beings of other types!

I said that the level of Lake Erie was once at the top of the heights of Lewiston, thirty-eight feet above its present altitude. This elevation submerged the flats to the east and west of the Detroit and St. Clair Rivers, and united Lake Erie with Lake Huron by a shallow expanse of water, which in some places possessed a breadth of fifty miles or more. Still farther, the level of Lakes Huron and Michigan was raised twenty-five feet above their present altitude, and a portion of the waters of the upper lakes found an outlet from Lake Michigan into the Des Plaines River, and thence into the Illinois and the Mississippi—if, indeed, a large portion of the prairie region of Illinois was not submerged by such an altitude of the lakes. At the same time, Saginaw Bay of Lake Huron stretched into the centre of the peninsula of Michigan.

This is not the highest altitude at which the waters of the lakes have stood, though the barriers which dammed them have long since disappeared. Along the southern borders of Lakes Erie and Ontario, the rocks arise from their more southern depressions, and face the lakes in bold escarpments three hundred and fifty feet above the respective levels of the waters. These bluffs have been the rocky shores of the lower lakes. For unnumbered ages the furious north wind has rolled mad waves against those adamantine walls, and battlement after battlement has tumbled down and been ground to powder by the tireless beating of the stormy surge. Between the foot of the mural escarpment and the present margin of the lake is a series of parallel terraces, each showing the altitude at which the receding waters have made a pause. These terraces along the southern shore of Lake Erie range from ninety to one hundred and twenty feet above the present level of the water. In Huron and Sandusky Counties the subsidence of the escarpment permits them to diverge a greater distance from the lake. Curving northward into Michigan, they pass through Monroe, Wayne, and Oakland Counties, and continue in that direction. They are recognized again on the shores of Lake Michigan.

At Mackinac Island are deeply engraved records of a higher level of the waters. The island itself is but a vestige of an ancient formation which once filled the straits, and joined the highlands on the west and south. It towers, a monument of the destructive agencies of geogony, three hundred and eighteen feet above the lake. The principal plateau of the island is one hundred and fifty feet above the lake. In the midst of this rises "Sugar Loaf," one hundred and thirty-four feet higher (Fig. 85). This is a remnant of the rocks forming the upper plateau which once extended over the whole island. It is a conical mass of brecciated limestone, all whose sides bear the marks of the long-continued action of the waves. On one side an ancient grotto has thus been exca-

Fig. 85. View of Sugar Loaf, Mackinac Island.

vated, into which the surges have rolled with the deafening reverberations of a sea-coast "purgatory." The principal plateau of the island is limited at nearly all points by an abrupt wall dropping down into the deep waters of the lake. Here is a beetling cliff, one hundred and forty-five feet high, called "Lover's Leap," connected with a sentimental Indian legend. In another place is "Chimney Rock," one hundred and thirty-one feet high, and in another, "Robinson's Folly." At "Arched Rock" (Fig. 86), on the eastern side, towering one hundred and forty feet above the lake, the fierce waves, unable to reach the solid and unyielding brow of the precipice, have mined beneath it, perforating the limestone wall; and a natural bridge hangs there, with one end resting on a winged abutment stretching toward the lake. All round the walls of this castellated and charming isl- and the recording waves have left their hieroglyphs, from the water's edge to the battle-

Fig. 86. Arched Rock, Mackinac Island.

ments, and he who can read the language may ponder there the vicissitudes of the ages.

While, during the high tides of the lakes, the erosive waves were gnawing at the rocks of Mackinac and Ohio, the waters of Lake Michigan, in a quieter mood, were performing a work equally enduring and peculiar. Here we find our attention challenged by the question of prairie origin and prairie features, but the views to be presented will be held in abeyance until a chapter on a subsidiary topic shall have been interposed.

CHAPTER XXIII.

VITALITY OF BURIED VEGETABLE GERMS.

I VENTURE here to enunciate a view which to many may appear incredible. For some years past I have been inclined to believe that the germs of vegetation which flourished upon our continent previous to the reign of ice, and many of which must have been buried from twenty to one hundred feet beneath the surface of the glacial rubbish, may have retained their vitality for thousands of years, or even to the present time. There are not a few indications that vegetable germs are capable of such preservation, and not a few that they thus exist in the ancient drift. The consequences of such remarkable preservation possess a geological importance so novel and interesting that I am sure the reader will be pleased with a view of the facts bearing upon the doctrine.

Many familiar facts may be cited which certainly have a significance far greater than has been generally suspected, and which tend to show that the seeds of vegetation are reposing in a dormant state in our ordinary soils and subsoils. Nothing is a more common observation than to see plants making their appearance in situations where the same species was previously unknown, or for a long time unknown, and under circumstances such that the supposition of a recent distribution of seeds is quite precluded.

The sudden appearance of unwonted species frequently occurs when a change is produced in the physical condition of the soil. Left to Nature, certain perennial grasses secure almost exclusive foothold in our fields, and form a sod in which the ordinary annuals are unable to flourish. Break up the sod after any number of years, and subdue the perennial grasses, and we shall have a crop of annuals the first season—Veronicas, Chenopodiums, Euphorbias, Portulacas, Ambrosias, Crab-grasses, Foxtails, Panicums, etc. Cease cultivation, and the Poas and Glycerias will immediately resume possession. Similarly, the pertinacity with which the common knot-grass seizes and maintains its position only along the hardest-beaten footpaths is notorious, while the greater plantain renders itself no less conspicuous growing alongside. Earth thrown out of cellars and wells is generally known to send up a ready crop of weeds, and, not unfrequently, of species previously unknown in that spot. In all these cases, after allowing for all known possibilities of the distribution of seeds by winds, birds, and waters, it still seems probable that germs must have previously existed in the soil.

Similar sudden appearances of new forms take place when a change is effected in the chemical nature of the soil. Illustrations are familiar to every agriculturist. How soon does a dressing of undecomposed muck, or peat, or sawdust develop a crop of acid-loving sorrel, and how readily is it again repressed by a dressing of some alkaline manure? Let the waters of a brine-well saturate a meadow, and how long before we witness the appearance of the maritime *Scirpus* and *Triglochin*, or some other salt-loving plant whose germs, unless spontaneously developed, must have lain dormant at a greater or less depth?

Something of the same nature is witnessed on the disappearance of dominant species, whether through the agency of man or the processes of Nature. It is well known that the clearing of a piece of forest and the burning of the brush is almost always followed by the appearance of certain unwonted plants known as "fire-weeds." In many cases it would seem highly improbable that the seeds of such plants had recently been transported to such situations at the moment when the disappearing forest admits the introduction of the conditions essential to their growth. It can hardly be doubted that the seeds existed in the soil, ready to germinate whenever free sunlight, warmth, and atmospheric air should be permitted to rouse their latent vital energy. Of the same nature is the recurrence of particular forest growths upon the same soil. Not unfrequently the second growth is of a very different nature from the first. In the "old fields" of Virginia and other Southern States, the soil, cleared originally of the deciduous forest, and then abandoned after years of continuous cropping, sends up a growth of pines instead of deciduous trees. In some parts of Southern Ohio, as I have been informed, a forest of unmixed locust-trees follows the destruction of the ordinary mixed forest.

Mr. Marsh, in his learned work entitled "Man and Nature," has quoted from Dwight's "Travels" his account of the appearance of a fine growth of hickory on lands in Vermont which had been permitted to lie waste, when no such trees were known in the primitive forest within a distance of fifty miles. He quotes also Dr. Dwight's account of the appearance of a field of white pines, on suspension of cultivation, in the midst of a region where the native growth was *exclusively* of angiospermous trees. "The fact that these white pines covered the field exactly, so as to preserve both its extent and figure," says Dr. Dwight, "and that there were none in the neighborhood, are decisive proofs that cultivation brought up the seeds of a former forest within the limits of vegetation, and gave them an opportunity to germinate."

In this connection may be quoted a statement of Darwin, in "The Origin of Species," to the effect that in the midst of a very sterile heath in Staffordshire, some hundreds of acres were planted with Scotch fir, and, after twenty-five years, not less than twelve species of plants (not counting grasses and sedges) had made their appearance in the plantation of firs, "which could not be found in the heath," and this though the fir-forest seems to have been visited only by insectivorous birds.

The existence of a succession of forests of different prevailing species has been satisfactorily established in Denmark by the researches of Steenstrup on the *Skovmose*, or forest-bogs of that country. These bogs are from twenty to thirty feet in depth, and the remains of forest trees in successive layers prove that there have been three distinct periods of arborescent vegetation in Denmark—first, a period of the pine; secondly, a period of the oak; lastly, a period of the beech, not yet arrived at its culmination. The dominant species of each period flourished to the entire exclusion of the other two species. Cæsar affirms that the *Fagus* (beech) and *Abies* (fir) were, in his time, wanting in England; but the beech is now plentiful; and Harrison tells us, in his "*Historicall Description of the Iland of Britaine*," that "a great store of firre" is found lying "at their whole lengths" in the "fens and marises" of Lancashire and other counties, where not even bushes grew in his time. No doubt such extinct forests have flourished in America, even since the Glacial Epoch, and have stocked the accumulating soils with their stores of vitalized fruitage, awaiting some future resurrection; and no doubt the "fens and marises" of Lancashire, under suitable circumstances, would reproduce from their granaries of forest fruit the arboreal vegetation which had flourished and disappeared before the Roman Conquest.

Mr. Marsh, in the work already quoted, after expressing his opinion that the vitality of seeds "seems almost imperishable while they remain in the situation in which Nature deposits them," proceeds to cite numerous instances in which one crop of plants has disappeared on a change of conditions, and another, of different nature, has promptly assumed its place, originating evidently from seeds preexisting for ages in the soil. He says, "Earth brought up from wells or other excavations soon produces a harvest of plants often very unlike those of the local flora." He goes so far as to express the opinion that earth ejected from considerable depths by a certain earthquake convulsion, to which reference is made, and which soon became covered with vegetation "never observed in that region before," must have brought with it the seeds from which the novel vegetation sprang, under "the influence of air and sun, from depths where a previous convulsion had buried them ages before."

From such facts as those which have been cited, it seems to be proven that the seeds of plants may retain their vitality in the soil and subsoil at least for quite a number of years. The facts show that the germs exist in places where we have no knowledge of their introduction, and in places where they could not probably have been introduced during the human epoch. Whence come the germs of that vegetation which is every where springing up in situations to which recent seeds could not have been distributed? This question has agitated the mind of many an inquirer who would have shrunk from the solution which I venture to offer. Let us examine the facts.

The vegetation which characterized the close of the Tertiary Epoch was probably nearly identical with that existing at the present day under the same climatic conditions. Even in the older Tertiary Lignites we have, according to the investigations of Lesquereux and Newberry, the remains of plants belonging to the following American genera, viz.: Oak, Hickory, Poplar, Maple, Mulberry, Hornbeam, Box-elder, Laurel, Bay, Dogwood, Sumac, Olive, Buckthorn, Magnolia, Smilax, White Cedar, Sequoia, Cypress, and Sabal. These identifications have been made from scanty and defective material, and we may fairly presume that further investigations will greatly increase the number. Yet these plants, belonging probably to the earliest Cenozoic Epoch, show, according to Lesquereux, "the greatest affinity with species of our own time." From other beds of the middle or earlier Tertiary we have still other existing genera, such as Persimmon, Beech, Black Gum, Aristolochia, etc. The facts in our possession relative to the middle and later Tertiary Epochs show a most decided approximation to the existing flora. From a pleiocene deposit near Somerville, Tennessee, Lesquereux identified the following recent *species*, viz.: Carolina Laurel, Carolina Plum, Myrtle-leaved Oak, and Common Beech. From the chalky banks of the Mississippi River, near Columbus, Kentucky, a collection was made, of which all the species are recent, viz.: Live Oak, Dwarf Chestnut, Winged Elm, Gmelin's Planer-tree, Entire-leaved Prinos, New Jersey Tea, Pecan, Honey Locust, and Sweet Flag. It is true that Dr. D. D. Owen has assigned the deposit containing these remains to the Post-Tertiary Age; but their position is one hundred and twenty feet below the ferruginous sands containing the bones of the extinct sloth *Megalonyx Jeffersoni;* and, as the nature of these species is incompatible with such a climate as we universally associate with the Glacial Epoch, it is quite likely this assemblage of vegetable remains represents the general nature of the arboreal flora in existence near the close of the Tertiary Age.*

* Dr. Newberry has shown that even the Cretaceous flora of North America was very similar to that now existing.—*Amer. Jour. Sci. and Arts* [2], xxix., 215 *et seq.* See also Lesquereux's determinations.—*Amer. Jour. Sci. and Arts* [2], xlv., p. 104, and xlvii., p. 286.

Although our positive knowledge of the vegetation of the period immediately preceding the advent of the reign of ice is confessedly meagre, it is certain that all the facts in our possession point to close specific correspondence with the modern vegetation of the same regions, modified, certainly, by the fact that even in the latest Tertiary the climate was considerably warmer than in the same latitudes at the present day.

All the luxuriant vegetation which flourished at the close of the Tertiary was undoubtedly swept off by the events which characterized the reign of ice, and, as has been already stated, the ruins of this vegetation were entombed in the rocky *débris* created by the moving glacier. The drift deposits became the vast granary in which Nature preserved her store of seeds through the long rigors of a geological winter.

But what evidences have we that the seeds of plants are capable of retaining their vitality through a geological period?

The ordinary process of destruction of vegetable tissues is merely an oxydation of the carbon and hydrogen entering into their constitution. I seriously doubt whether the requisite conditions for such oxydation exist at considerable depths in the soil. Mr. Jabez Hayden, of Windsor Locks, Connecticut, has a small quantity of corn, which is part of a bushel or more uncovered by the breaking away of the banks of the Connecticut River, a little above the mouth of the Farmington, not many years since. It probably dates back prior to the settlement of Windsor in 1635. The kernels had been charred and buried below the ordinary depths of cultivation (*Stiles's Hist. Ancient Windsor*, p. 85).

It is stated that the piles sustaining the "London Bridge" have been driven five hundred years, and are still comparatively sound. Old Savoy Place, in the city of London, is sustained on piles driven more than six hundred and fifty years ago, and they are yet perfectly sound. The ancient and historic city of Venice consists of brick and stone structures resting upon wooden piles which were driven in the seventh and eighth centuries. One of the piles taken up from the bridge built by the Emperor Trajan across the Danube was found petrified to the depth of three quarters of an inch, while the remainder of the substance was unchanged after an interval of sixteen hundred years. The timber maul-handles, shovels, and other wooden implements found in the ancient mines of Lake Superior still remain in a good state of preservation in cases where they have been immersed in water; and the wheels employed in draining some of the ancient Roman mines in Spain are represented to be in a perfect state of preservation after the lapse of fourteen hundred and fifty years. The ancient piles in the lake habitations of Central Europe retain a remarkable degree of soundness, though driven before the epoch of written history.

Passing beyond the range of human records, we remark the existence, along the Atlantic borders of New Jersey, of extensive buried swamps, in which the trunks of the white cedar (*Cupressus thyoides*, not the "White Cedar" of the West) are found in such a state of preservation that the inhabitants work them up for lumber. So extensive are these deposits of buried tree-trunks that the "mining of timber" has long been a prominent branch of business along some parts of the beach (Fig. 87). They lie from two to fifteen feet beneath the surface. We may form some conjecture in reference to the antiquity of these fossil cedar swamps from the age of the trees which have evidently grown upon spots that had been occupied by still earlier generations of trees. Professor Cook informs us

Fig. 87. "Mining" Cedar Logs in an ancient Buried Swamp on the coast of New Jersey.

that the number of annual rings in the trunk of one of these buried trees six feet in diameter was one thousand and eighty, while beneath it was another trunk counting five hundred rings, which had evidently grown and fallen down before the superincumbent tree had commenced its growth. In other instances, the relative positions of trees and stumps are such that we are compelled to assign to perfectly sound timber, retaining even its characteristic aromatic odor, an antiquity of hundreds and even of thousands of years. (See Cook, Geology of New Jersey, 1868, p. 343, etc.; Lyell, Second Visit to the United States, vol. i., p. 34.)

Buried tree-trunks are often exhumed from glacial drift at the depth of twenty to sixty feet from the surface. Dr. Locke has published an account of a mass of buried drift-wood at Salem, Ohio, fifteen miles north of Dayton, where it lies from thirty-seven to forty-three feet beneath the surface, imbedded in a layer of ancient mud. The museum of the University of Michigan contains several fragments of well-preserved tree-trunks exhumed from wells in the vicinity of Ann Arbor. Such occurrences are by no means uncommon. The encroachments of the waves upon the shores of the "great lakes" reveal whole forests of the buried trunks of the White Cedar (*Thuja occidentalis*), bearing scarcely a trace of the work of destructive agencies upon them.

Unaltered vegetable structures have been found in geological deposits of even higher antiquity. It is known that well-preserved woody tissue has been frequently exhumed from deposits of Tertiary, and even of greater age. I am in possession of pieces of drift-wood from the Cretaceous sands of Alabama, in which the ligneous tissue is so fully preserved as to be capable of ignition, like recent wood. Even from the Coal Measures of Michigan I have made preparations of the delicate tissues of the fronds of so-called Scale-mosses (*Jungermauniaceæ*); and from the coal mines of Lasalle, in Illinois, I have collected specimens of exogenous wood of a brown color and not yet carbonized, though partially pyritized. All these examples tend to show the extreme slowness of the process of decay in ordinary vegetable tissues when excluded from the usual conditions of decay by burial in the earth.

The oily tissues of which seeds are composed are still more capable of resisting the tendency to dissolution, and ought certainly to remain unchanged, under circumstances which permit such perfect preservation of ordinary ligneous fibre. The evidences are very conclusive that the seeds of ordinary vegetation may lie dormant in the surface-soil for half a dozen or a dozen years. The seeds of the various "fire-weeds" which spring up on a forest clearing after the brush has been burned off, must have reposed in a latent state during the existence of the forest whose disappearance is the signal for the resumption of their vital activity. The same is true of the seeds of the "old field-pines," which have probably lain for an age or more, awaiting the maturity and destruction of the deciduous forest which usurped the soil. How *many* ages may they have lain there? How many more might they have lain, and still been found ready for the first opportunity to seize a foothold?

There are some facts in our possession which are still more specific. It is well known that Dr. Lindley raised three raspberry plants from seeds discovered in the stomach of a man whose skeleton was found thirty feet below the surface of the earth, at the bottom of a barrow or burial-mound which was opened near Dorchester, England. With the body had been buried some coins of the Emperor Hadrian, from which we are justified in assuming that these seeds had retained their vitality for the space of sixteen or seventeen hundred years. If they remained undamaged that length of time, their condition was practically fixed; and who shall say that ten thousand years would have produced a greater effect?

Again: Lord Lindsay states that, in the course of his wanderings amid the pyramids of Egypt, he stumbled on a mummy proved by its hieroglyphs to be at least two thousand years of age. On examining the mummy after it was unwrapped, he found in one of its closed hands a bulb, which, when planted in a suitable situation, grew and bloomed in a beautiful dahlia. The credibility of this story is very questionable, since the real dahlia is a tuberous-rooted Mexican genus, not known to botanists till the year 1789. That a bulb of any sort germinated under the circumstances alleged is highly improbable, since the characteristic of the surroundings of a mummy is perfect dryness, which would completely change and devitalize the tissues of a bud-like bulb. It is, however, more credibly asserted, and generally believed, that wheat is now growing in England which was derived from grains folded in the wrappings of Egyptian mummies, where they must have lain for two or three thousand years. Professor Gray, the eminent American botanist, does not fully credit the account, but Dr. Carpenter, the distinguished English physiologist and naturalist, gives it his full indorsement.*

Professor Agassiz asserts that "there are some well-authenticated cases in which wheat taken from the ancient catacombs of Egypt has been made to sprout and grow." Dr. Carpenter even goes so far in this connection as to give utterance to the following observations, which happen to be extremely pertinent in the present instance:

"These facts make it evident," he says, "that there is really no limit to the duration of this condition (latent vitality), and that when a seed has been preserved for ten years, it may be for a hundred, a thousand, or ten thou-

* On this subject and the longevity of seeds in general, see Report of the Commissioner of Patents for 1857, Agriculture, p. 256 (condensed from the Gardener's Chronicle, London).

sand, provided no change of circumstances either exposes it to decay or calls its vital properties into activity. Hence, where seeds have been buried deep in the earth, not by human agency, but by some geological change, it is impossible to say how long anteriorly to the creation of man they may have been produced and buried, as in the following curious instance: Some well-diggers in a town on the Penobscot River, in the State of Maine, about forty miles from the sea, came, at a depth of about twenty feet, upon a stratum of sand. This strongly excited their curiosity and interest, from the circumstance that no similar sand was to be found any where in the neighborhood, and that none like it was nearer than the sea-beach. As it was drawn up from the well it was placed in a pile by itself, an unwillingness having been felt to mix it with the stones and gravel which were also drawn up. But when the work was about to be finished, and the pile of stones and gravel to be removed, it was necessary also to remove the sand-heap. This, therefore, was scattered about the spot on which it had been formed, and was for some time scarcely remembered. In a year or two, however, it was perceived that a number of small trees had sprung from the ground over which the heap of sand had been strewn. These trees became, in their turn, objects of strong interest, and care was taken that no injury should come to them. At length it was ascertained that they were Beach-plum-trees; and they actually bore the Beach-plum, which had never been seen except immediately upon the sea-shore. The trees had therefore sprung from seeds which were in the stratum of sea-sand that had been pierced by the well-diggers." It can not be doubted, as Carpenter concludes, that the seeds of the Beach-plum had lain buried since the remote period when that part of the state was the shore of the slowly-receding sea.

Such a fact, so striking and so circumstantially recorded, is only of the same nature as others less critically noted, which daily pass before our eyes in the upspringing of vegetable forms from the diluvial materials thrown out of wells, cellars, and other excavations.

The bones, the hair, and even the flesh of the extinct mammoth have been preserved in glacial deposits on the shores of Siberia. In so complete a state of preservation has the flesh been found, that dogs and bears greedily devoured it. If a material so perishable as muscular fibre could be preserved since an epoch which antedates authentic history, is it not more probable that the oily tissues of vegetable seeds could resist the tendency to decay under similar circumstances?

It must be confessed that the crucial observation is yet to be made. If vegetable germs exist in the drift, they can be discovered beforehand. I am not aware that any thorough search has ever been made for them; but, until they have been actually detected, it is probable that even the convincing facts cited above will fail to secure universal assent to the doctrine of the prolonged vitality of the seeds of pre-glacial vegetation. While, however, the case is far from demonstrated, it may fairly be submitted that the explanation of certain facts afforded by this theory is less presumptuous and improbable than the supposition of spontaneous generation, the fortuitous distribution of seeds by any modern agency, or any other explanation that has yet been offered.

CHAPTER XXIV.

PRAIRIES AND THEIR TREELESSNESS.

THE prairies of the Mississippi Valley, especially those lying within the limits of the great State of Illinois, constitute one of the most remarkable features of North American topography. Hundreds of thousands of acres, stretching through all the central and western portions of the state, present a scene of almost unbroken level and treelessness. The great prairies are neither a perfect plain, nor in all cases completely undiversified with arboreal vegetation. The surface is generally undulating; and here and there rise gravelly knolls and ridges on which the timber has obtained a foothold. But these wooded spots are often many miles apart, and scarcely serve to rest the eye, wearied with the monotony of an interminable clearing, fenceless meadows, and unsheltered farm-houses.

The traveler, leaving Chicago by one of the great southern routes—for instance, the Chicago, Alton, and St. Louis Railway—passes out through the muddy and straggling outskirts of the Western metropolis, and, ere he has thought of the great prairies through which he had expected to pass, he finds himself at sea. Looking from his car-window, the country landscape seems at first to be entirely wanting. One feels as if passing over a trellis-bridge three hundred feet above the surrounding region. The customary objects—forests, shade-trees, fences, houses, distant hills—which elsewhere lift themselves to the horizontal plane of the eye, are not here. The traveler must make the second effort, and look down upon the level of the country upon whose bosom he has launched. The sensation is that which one experiences in going to sea. The rattling train is easily transformed into the puffing and creaking steam-ship, while the interminable prairie, mingling its distant and softened green with the subdued azure of the summer sky, can be likened to nothing but the ocean's boundless expanse. The ever-recurring undulation of the prairie is the grand ocean-swell which utters perpetually a reminiscence of the last storm, while the evening sun, with dimmed lustre, settles down into the prairie's green sod, as to the mariner he sinks into the emerald bosom of the sea.

Illinois has been styled the garden state of the West. The deep, rich, pulverulent soil of the upland prairie, and especially its readiness for the plow, without the intervention of a year's hard labor in opening a clearing, have always constituted powerful attractions for the settler from the stony soils of New England, and the wooded regions of the other states. It is extremely doubtful, however, whether the absence of forests over the area of half a state possesses a balance of advantages. Forests possess immense utilities in addition to furnishing lumber and fuel. This discovery was long since made in the denuded regions of the older European countries; and Americans are talking at times as if they were growing wiser. Even the cobble-stones of a New England or New York soil are not unmitigated inconveniences. During the day they absorb the warmth of the sun, and at night they retain it and impart it to the soil. In times of drought they screen the soil from the direct rays of the sun, and thus moderate the intensity of the heat. They diminish the evaporating surface of the soil, and thus diminish the effects of continued droughts. A loose stone is a shade; but, unlike a tree, it has no roots of its own to creep about and steal the moisture from weaker forms of vegetation. A few stones do not diminish materially the amount of soil upon an acre; and, with the benefits which they confer, it is doubtful whether they are not actually to be desired, especially in regions subject to drought. A field will produce no more grain with the stones picked out than with the stones left in.

From our earliest knowledge of the prairies, speculation has been rife as to their origin. The old and popular belief was that which attributed their treelessness to the annual burning of the grass by the Indians. But the prairies present other phenomena which the annual burning fails to explain. Besides, the treelessness remains in regions where the burnings have ceased. And, lastly, the treeless prairies were not the only regions burned by the Indians. And if they were, it seems more likely that the Indian burned the rank grasses because the region was treeless, than that the region became treeless from the burning of such vegetation as flourishes in the shade of a forest.

It has sometimes been suggested that the region was originally forest-covered, and that the southern cane flourished in such luxuriance amongst the trees as to rob them of their moisture and nourishment, and thus cause their extinction. The cane, having deprived itself of the protecting shade of the forest, was in turn scorched out by the rays of the summer sun. This theory is every way unsatisfactory.

With others, the absence of trees is to be attributed to the dryness of the atmosphere—and consequently of the soil—at certain seasons of the year. It can not be doubted that the treeless plains of the far West, and also other regions, have failed to produce arboreal growths through an insufficient supply of moisture. Still other treeless regions are such from an excess of saline constituents in the soil. But all such regions have nothing in common with the prairies of Illinois except their treelessness. The topography and soil-constitution of the Illinois prairies points to a different and a peculiar history. Moreover, trees occupy the drier knolls of the prairies in the midst of the common atmospheric conditions.

Exactly the reverse of this theory is that which attributes the absence of trees to an excess of moisture in the soil at certain seasons. But we well know that there is no soil or situation so wet and stagnant but certain trees will flourish upon it—the willow, the cottonwood, the beech, the black ash, the alder, the cypress, the tupelo, the water-oak, the tamarack, the American arbor-vitæ, or some other tree—some of them standing joyously half the year, if need be, in stagnant water. Many swales are indeed treeless; but is this in consequence of the inability of a willow to take root and maintain itself, or rather in consequence of the formation of the swale in times so recent that the germs of trees have not yet been scattered over it? Moreover, wetness can not be attributed to many portions of the Illinois prairies which are entirely treeless. Is there a different cause for treelessness here?

Lastly, it has been suggested within a few years, by high geological authority, that the lack of trees is caused by excessive fineness of the prairie soil. It can scarcely be denied, however, that other soils, as pulverulent as that of the prairies, are densely covered with forest vegetation, and that in the same latitudes and under the same meteorological conditions. On the other hand, certain soils of a coarser texture are equally treeless. But the final objection to this theory, and to all theories which look to the physical or chemical condition of the soil, or even to climatic peculiarities, for an explanation of the treeless character of the upland prairies of the Mississippi Valley, is discovered in the fact that *trees will grow* on them when once introduced—not water-loving trees exclusively, but evergreens, deciduous forest-trees, and fruit-trees, such as flourish in all the arable and habitable portions of our country. Every one will now admit that trees flourish upon the prairies. In proof of the fact, the prairie farmers are actively engaged in their introduction. "The prairies * * *," says Gerhard,* "may be easily converted into wooded land by destroying with the plow the tough sward which has formed itself on them. There are large tracts of country where, a number of years ago, the farmers mowed their hay, that are now covered with a forest of young, rapidly-growing timber. * * * A resident of Adams County testifies to the effect that locust-trees planted, or, rather, sown on prairie land near Quincy, attained in four years a height of twenty-five feet, and their trunks a diameter of from four to five inches. * * * In like manner, the uplands of St. Louis County, Missouri, which were in 1823 principally prairie lands, are now covered with a growth of fine and thrifty timber, so that it would be difficult to find an acre of prairie in the county." This testimony is confirmed by numbers of persons from various parts of the state whom I have questioned on the subject. The introduction of timber as a branch of rural industry is now systematically pursued. The principal drawback to the cultivation of forests and fruit-trees is the violence of the prairie winds and the occasional severity of the wintry weather.

If what I have suggested in reference to the persistent vitality of buried vegetable germs be true, we have a ready, simple, and beautiful solution of this long-vexed problem.

There are pretty satisfactory evidences that the soil of the prairies is of lacustrine origin. It has the fineness, color,

* Illinois as it Is, p. 277. Compare also Wells's *Amer. Jour. Sci. and Arts*, i., 331; Engelmann, *Ibid.* [2], xxxvi., 389; Edwards's *Rept. Dept. of Agric.*, 1862, p. 495.

and vegetable constituents of a soil accumulated upon a lake-bottom. We find in it, moreover, abundant fossil remains of a lacustrine character. Fresh-water shells of species still existing in Lake Michigan are found in localities many miles from the existing shore. Finally, we have found all around the chain of the great lakes abundant proofs that their waters once occupied a much higher level than at present. We have discovered the obstacle which dammed the waters to this extraordinary height. In short, we have ascertained that the prairie region of Illinois *must have been* a long time inundated, whether such inundation contributed to the characteristics of the prairies or not. I think it did. If I ascertain that the cause for an inundation exists; if I see the traces of an inundation all the way from Niagara River to Illinois; if the barrier which shuts out Illinois from the lake is not one third the height of the ancient lake-flood; if I find throughout the region exposed to inundation the peculiar soil deposited by fresh waters, together with traces of lacustrine animals which never wander over land, do I not discover a chain of facts which necessitates my conclusion? During the flood-tide of the lakes, Lake Michigan must have found an outlet toward the south. We find corroboration of this. The broad, and deep, and bluff-lined valley of the Illinois River was never excavated by the present inconsiderable stream. The deserted river valley discoverable at intervals farther north, indicates the former southward flow of a large volume of water. At Lemont this valley is distinct, with its bounding bluffs, and its "pot-holes" worn in the solid rock of the ancient river-bed. This was the work of the lake in its declining stages. At the earlier period, when the waters of Lake Michigan stood one or two hundred feet above their present level, how much of the region south and west of Chicago must have been submerged?

The ancient lake must have reached its arms into Iowa, Northern Indiana, and Southwestern Michigan.

While the expanse of lacustrine waters was brooding over the region destined to become a prairie, they busied themselves in strewing over the tombs of pre-glacial germs a bed of mud which should forever prevent a resurrection. Lake sediments themselves inclose no living germs. You will see the seeds of grasses and the fruits of trees, washed in by the recent storm, floating upon the surface, and eventually drifting to the lee-shore. If they ever sink to the bottom, and wrap themselves in the accumulating mud, it is after they have lost their vitality. Sunken and buried, they go to decay. Let a lake be drained, and the bottom remains a naked, barren, parching, shrinking waste. No herbs, or grasses, or trees burst up through the pottery-like surface. But every where, from beds of ancient glacial materials, vegetation is bursting forth and announcing itself. Lo, here I am! speaks the nodding young pine that had been slumbering just beneath the surface through the long and undisputed possession of the deciduous forest which the axe has just mown down. Not so in a lake-bottom. Here are the cerements of the dead, not the wrappings of the slumbering.

When, therefore, the ancient lake relinquished dominion over Central Illinois, he left a devastated and desolate country. Around the ancient shores of the abandoned area the emerald forest had stood nodding, and blossoming, and fruiting, while the inundating lake had washed the slopes down which the oaken and beechen roots descended to sip refreshing draughts. Ever since the time when the Atlantic and Pacific last held carnival in the Mississippi Valley, these vigorous trees had stood smiling upon the face of the freshening residuum left in Illinois on the final retreat of the oceans. A resurrected forest had risen from the tombs of the preceding epoch. And not alone around the borders of the widened lake, but upon every island knoll which raised its head above the denuding waters. This encircling forest and these isolated island clumps still stood and flourished when at length the lake receded.

No turf carpeted the abandoned lake-bottom. No oak, or beech, or pine raised its head through the covering of lake-slime which separated the slumbering-place of vegetable germs from the animating influence of sun and air. By degrees, however, the floods washed down the seeds of grasses and herbs upon the desert area, and humbler forms of vegetation crept from the borders toward the centre. At length the entire area smiled with vernal flowers, and browned in the frosty blasts of winter. The bulky acorn, and walnut, and hickory-nut traveled with less facility, and the forest more sluggishly encroached upon the lake's abandoned domain. In this stage of the history the Indian was here. For aught I know, he was here while yet the prairies were a lake-bottom. His canoe may have been paddled over the future spires of Bloomington and Springfield, and the muscalonge may have been pursued through the future streets of Chicago; but, at least, the Indian was present in the interval of time by which the herb distanced the tree in their race for possession of the new soil. In this interval he plied the firebrand in the brown sedges of autumn, and made for himself an Indian-summer sky, while he cleared his favorite hunting-ground of the rank growths which impeded both eye and foot. While the Indian was engaged in these pursuits, and while yet the forest had not had time to extend itself over the prairie, the white man came up the lake from Mackinac, crossed over the prairies to the Mississippi, saw the Indian engaged in his burnings, and hastily concluded that this was the means by which the trees had been swept off—ignorant of the history that had passed, and which was even then, as now, in very progress, and which was even then, as now, actually crowding the forest upon the prairie, and bringing about the day when, perhaps a thousand years hence, the prairies, like the forests of Lancashire, will live only in history.

CHAPTER XXV.

SOMETHING ABOUT OIL.

THE very word has wrought like magic. The smell of the article has turned men crazy. It has opened purse-strings which the cries of the orphaned, the tears of the widowed, and the pleas of religion could never loose. It has made men lavish in a hopeless enterprise who had no pence to spare under the counsels of wisdom. It has caused men to scorn the admonitions of the instructed and professional, to trust their own stark ignorance in the stake of a fortune. It has led the self-reliant and pursey capitalist to heap contempt on the wisdom and experience of science, to follow the lead of his own olfactory. All this because "oil" is a synonym for gold.

Auri sacra fames! quid non mortalia pectora cogis?

Since the historical excitement of the "South-Sea Bubble," the business world has hardly been invaded by such a fever of speculation as raged over the Northern United States from 1862 to 1866. When it was positively settled that oil could be drawn from the solid rocks—oil suited to the uses of illumination, gas-making, fuel, and lubrication—men who have the keenest eye to utility, and who counterpoise all values with bullion, were constrained to admit that Providence had done more for our race than they had ever dreamed. No doubt many men made suitable recognition of the services of the Almighty in facilitating the ends of money-getting. The picture which memory treasures, however, is that of a herd of porcine quadrupeds jos-

Fig. 88. The Noble Oil Well, Penn.

tling each other for the largest share of their master's allowance.

At first it was generally supposed that one locality was as likely as another to yield the oleaginous fluid, and experiments innumerable were instituted wherever men could be found whom the infectious fever had reached. We now know that not one neighborhood in a thousand affords the geological conditions requisite to success. Another precipitate and erroneous conclusion was that which assumed the surface configuration of the earth to be the only essential condition of oil accumulation. Wherever a region could be found with a physical geography like that of Venango County—wherever a creek like Oil Creek had scored a country underlaid by sandstone like Northwestern Pennsylvania—there might have been seen the men whose experienced olfactories were employed to test the odor of every bog, and stain, and film which prying eyes could bring to light. Especially if such a creek were bordered by a flat walled in by rocky bluffs—but most especially if such a flat could be found at the fork of two streams, environed by rocks and hills of Pennsylvania sandstone, were the "oil-smellers" in high ecstasies. Happy the squatter whose steep and rugged hill-sides and narrow intervales afforded these first-class evidences of "productive property." I know of many an instance in which his land was tripled in market value by the magic touch of the magician of the hazel wand. The same kind of sandstone was essential; and it is marvelous that Nature had so disposed it that the oil-seeker could in every instance detect also the "first," "second," and "third" sandstones after the Venango style. No matter upon what formation the exploration might be progressing—perhaps a thousand feet below or above the geological horizon of Venango County—these oil-hunters, who had a wisdom above geology, could infallibly parallelize every formation with that of Venango County.

Another popular error was that of regarding beds of coal as the source of the oil. This led searchers for the coveted fluid to prefer the borders of coal-fields, or even the regions underlaid by coal. Often it seemed to be a matter of indifference whether it were calculated that the oil would naturally rise or sink through the rocks. With many the question was never considered. With most, however, the opinion was entertained—and to this day is cherished—that oil naturally descends through the strata. I have seen it gravely stated in published treatises on the subject that our native petroleum is the "drainage of the coal-measures." Nothing could be more erroneous. What connection can exist between the oil deposits of Enniskillen (Ontario) and the nearest coal-beds, at least one hundred miles removed? What between the oil accumulation of Manitoulin Island and the nearest coal-beds two hundred miles distant? Moreover, the coal-measures are every where less saturated with oil than many formations of more ancient origin.

"Surface shows" have been the fascination of many. The places of most copious escape to the surface were regarded as the favored spots where the "drainage from the coal-measures," in disregard of the laws of gravity and hydrodynamics, had obligingly deposited itself. Such "shows" were always illusory. A great "surface show" is a great waste. When Nature plays the spendthrift she retains but little treasure in her coffers. This was the lesson learned at great cost by the confident capitalists who took "stock" in the "surface shows" of Paint Creek, in Northeastern Kentucky. The production of petroleum in quantities of economical importance has always been from reservoirs in which Nature for ages had been hoarding it up, instead of making a superficial and deceptive display of her wealth.

But there is no illusion that has levied a heavier tax upon the folly and credulity of oil-seekers than the bituminous smell emitted by certain rocks, accompanied generally by the visible presence of more or less of the bituminous matter. The Corniferous limestone must take the praise of being the most successful fool-detective of any touchstone ever applied to the herd of oil-seekers. This limestone is very remarkable for the general abundance of oily and bituminous matters disseminated through it. Not unfrequently fragments of the rock present a black color and unmistakable pitchy smell, which are quite seductive. I have often seen it dripping with a tarry exudation which could be gathered up. Nay, I have seen clear petroleum flowing from small cavities in the formation, and published statements of the phenomenon as early as 1859 and 1860. Nevertheless, the conditions of oil-accumulation are not fulfilled in this formation, and, as a historical fact, no productive well is known to be supplied from the formation.

These observations were made by certain geologists while yet the oil excitement continued at its height. They were published to the world as warnings against the deceptive solicitations to investment which this formation presented. And yet men daily suffered themselves to be deceived. A thousand dollars spent in drilling a hole in this limestone was cheerfully paid by men who could not be persuaded to offer five cents for the endowment of a chair of geology in some reputable college or university. Hundreds of such holes were bored. The business had the mysterious fascinations of a lottery. The failure of a company at Toledo did not deter from an identical venture in Sandusky. A dozen failures in the neighborhood of London, Ontario, did not suffice to insure from the enchantment the very next man who gazed upon a limestone cliff reeking with the oozings of bitumen. And all this infatuation was indulged in spite of scientific advice, or in blissful ignorance of scientific teachings. Sweet is anticipation. An Ohio man showed me one day a quantity of fragments of this limestone, which were completely saturated after the usual style. It was a new sight to him, and he felt assured that Nature had simply used them as a roofing over an immense reservoir of oil. I recognized the formation at a glance, and remembered fifty instances in which it had been pierced without success. I assured the gentleman that it would be useless to bore in that rock. My advice saved a friend from becoming a fellow-victim, but the Ohio gentleman returned, and, like hundreds of others, resolved to trust his own ignorance in preference to professional skill. He bored his hole, and— it is still there!

In another instance, a gentleman of another state became fascinated by the smell of oil about an old stone-quarry in the Corniferous limestone. "Surely the oil must be treasured in these rocks," he said to himself. So, at great expense, he leased ground, erected buildings, employed hands, and bored a hole about six hundred feet in depth. As in all explorations of this formation, the never-failing smell of oil was continually taken for a "fine show," and he persevered in pushing downward. At length, however, the smell of oil gave out, and courage was kept up by smelling occasionally a piece of the surface-rock, or stirring the mud and water that had accumulated in the depressions of the quarry. The smell was a perpetual invigorator. Every sniff was worth fifty dollars to the grand enterprise. Every gas-bubble that could be conjured to the surface was good for another check. But at length the enterprise had been prosecuted to such an extent that it was regarded as just upon the eve of consummation and fruition. Now the geologist must be called in to puff the enterprise and sell the stock. Alas! he had the ungrateful duty of informing his employer that he had pierced *entirely through* the Corniferous limestone that had cheered him with its aromatic exhalations, and that he had entered the Niagara limestone, and would probably "strike fire" before he "struck oil." To palliate the disappointment, he had to add that the result could positively have been announced in the beginning, without the expenditure of a dollar in boring. This man expended six or eight thousand dollars in a most inexcusable and wasteful ignorance of the geological conditions, but yet endeavored to recover half of the moderate geological fee which he had paid to be informed of the hopelessness of his case.

The Corniferous limestone is extensively distributed throughout the West, and has afforded a wide field for the display of credulity that could not believe the truth, and avarice that could not spend enough in a bootless enterprise. It stretches in a broad belt from Columbus, in Ohio, northward to near the state line, where it bifurcates, one belt trending northwestward across Northern Indiana and into Southwestern Michigan, passing under Lake Michigan, and curving eastward, so as to reappear in the northern part of the state and form the headlands about Mackinac. The other belt trends northeastward, passing into Southeastern Michigan, beneath the western end of Lake Erie, and reappearing in the neighborhood of Woodstock and London, in Ontario, whence it deflects to the northwest, and passes under the middle of Lake Huron, reappearing in the headlands and islands of Michigan some distance southeast of Mackinac. Throughout nearly this whole extent it has been riddled by borers for oil, but to this day no productive and paying well has ever been opened in this formation.

The Niagara limestone has proved locally a similar cause of mistaken ventures. In the neighborhood of Chicago some of the beds of this limestone are eminently bituminous. Chicago has several times made discoveries which were destined to enrich her—if we could believe the newspaper accounts of the amount of fatness in her rocky substratum. Indeed, "the spirits" themselves, in looking down through the rocks which underlie the city, were egregiously humbugged by this rock "of color." Samples of this Chicago humbug may be examined in the old ivy-mantled and oil-bedraggled walls of the Second Presbyterian Church in that city. The oil fried out of the rock by the summer's sun has admirably imitated the dusky brush of antiquity, as modern art has learned to imitate the time-scarred products of the pencil of a Rubens or a Raphael.

The Coal-measures, also, from the wide-spread belief that they are the source of native petroleum, have been faithfully explored and expensively bored, with scarcely better success than in the Corniferous limestone. There has hardly been a good well that is known to have been supplied from the Coal-measures.

Crude petroleum is not a product of definite composition. It seems to be a varying mixture of several hydrocarbons, some of which, as naphtha, volatilize with rapidity when exposed to the atmosphere; others, as kerosene, slowly; while others, as bitumen, are nearly fixed. It contains also varying quantities of aluminous matters and other impurities.

It occurs in stratified rocks of all ages, from the Laurentian to the recent. It has even been observed in some rocks of a granitic structure. The mere presence of petroleum in a formation is far from being evidence that it exists in large quantities. Observation has shown that it does not exist in large quantities in any formation, except under certain intelligible conditions. Its presence in small quantities is to be expected.

It is an opinion almost universal among geologists that petroleum has been produced from organic remains. Hence, long before the discovery of the Eozoön in Laurentian rocks, it had been inferred that organic life existed upon our planet during the accumulation of those rocks, because, among other reasons, they afford conspicuous quantities of petroleum. Geologists are somewhat divided in opinion as to whether animal or vegetable organisms have afforded most of the native oil. Little dissent exists, however, from the doctrine that most of the oil occupying the pores and pockets of fossiliferous limestones has been derived from animal bodies, while that saturating shales and arising from shales has had a vegetable origin. As the oil of commerce is probably derived from the latter source, it appears that we are to regard our commercial oil as a vegetable product.

Petroleum and the other hydrocarbons are produced from organic matters by distillation in closed vessels. Any vegetable substance is capable of affording them. The refuse of the kitchen may be made to illuminate the mansion. Artificial distillation of any of the rocks containing organic remains gives rise to petroleum. Ordinary black shales abound in vegetable matter mostly in a state of comminution, and they readily afford large quantities of oil and gas. They are, in fact, distilled on the large scale in some European countries for the sake of these products. It has also been undertaken in this country, but without favorable results economically. Nature herself is engaged in this business, and competition with her is hazardous. Cannel coal, however, which is only a highly carbonaceous shale, was used for the production of "coal oil," or "Breckenridge oil," some years before the discovery of native petroleum. Hence arose the name "coal oil," which, in some sections of the country, is still applied to refined petroleum, although it never had any relations to the Coal-measures. Peat and lignite are capable of employment for the same purposes.

The deep-seated shales of the earth's crust are inclosed in rocky retorts hermetically sealed. The unquenched fires of the molten nucleus of the planet continue to impart their warmth to the ever-cooling crust. The rocky retorts in Nature's vast laboratory are warmed—their organic contents undergo a slow distillation—the products escape in the form of gas or oil, and slowly filter through pores and crevices toward the surface, till intercepted by some impervious stratum. These products, from the nature of the case, can not descend. They are lighter than water, and must tend to rise through the water in the midst of which they are disengaged.

The largest portion of oil and gas thus elaborated escapes to the surface and is lost. In order to prevent this escape, the retort must be furnished with a closed condenser or receiver. The exhalation ascending from the mother shale must be intercepted by a stratum of a clayey and impervious character. Beneath this the oil and gas will accumulate, displacing the water previously occupying the space. This reservoir may be an open cavity, a fissure, a shattered stratum of rock, or a mere porous sandstone. Here the oil will be stored.

But it is obvious that in the course of time a tendency will be manifest toward lateral extension over an indefinite distance, so that the products will be little concentrated in place, if they do not even find a leak in the roof and slyly escape to the surface. In order that these products may be locally restricted, the impervious stratum must present the form of a dome or roof. The underlying strata may, and generally do, conform in position to the roofing strata. We have here the requisite conditions for accumulation. Some portion of the oil and gas may filter through to the surface, or it may not. Obviously, if the outlet be large, the product must escape as fast as elaborated. If the reservoir be nearly closed, it may hold the products of the slow distillation of thousands of years. When one of these store-houses is exhausted it will be filled again, but perhaps not before the millennium.

I said that the oil and gas would displace the water previously occupying the spaces beneath the roof. It is plain that these substances must be hard pressed by the surrounding waters, re-enforced as they are on all sides by a virtual column reaching to the surface of the earth, which may be a hundred or five hundred feet above. The lateral pressure of a column of water five hundred feet high is enormous. All this the forming oil and gases must resist. No wonder that when given vent from above they sometimes burst forth with tremendous violence. At a well which I visited in Knox County, Ohio, the pressure of the confined gas was 180 pounds to the square inch, in addition to the pressure of a column of water 600 feet high. It escaped from the mouth with a roaring sound which could be heard at the distance of a mile. The supply was sufficient to illuminate a large city, and it continued to escape for several months.* When conducted horizontally through a pipe to the outside of the building and ignited, it formed a ragged and spiteful stream of fire of the diameter of a hogshead, which roared like a conflagration, and caused an illumination which was seen at the distance of sixteen

* This was in May, 1866. A letter from Peter Neff, Esq., of Gambier, dated June, 1868, states that this well is still "blowing."

miles. We can form but little conception of the circumstances under which such an enormous volume of gas can be confined at the depth of six hundred feet beneath the surface.

The escape of oil at the surface of a well is caused sometimes by mere hydrostatic pressure, as water rises in common Artesian wells. More frequently, perhaps, the oil is forced up by the elastic reaction of confined gases. An open cavity, or a porous portion of rock bounded on all sides by impervious walls—which constitutes a virtual cavity—may be partly filled with oil, while gases occupy the higher portions of the cavity. Such a cavity, whether actual or virtual, may possess any form or extent—or may consist of a number of cavities connected by narrow passages or mere fissures. In nearly all cases, more or less gas accompanies the oil, and subsists under a very high degree of pressure. The pressure in such cases is not the hydrostatic pressure of water, but a consequence of the continued generation of gas and oil long after the cavity had been filled. If a boring happens to penetrate the higher portion of such a cavity (Fig. 89), the gas at once rushes forth with greater or less violence and persistence. As soon, however, as the tension is relieved, the escape ceases. No oil will be obtained in such a case without applying suction, since there is no hydrostatic pressure exerted from behind, and the reaction of the gas tends rather to confine the oil in the lowest ramifications of the cavity.

Suppose, however, on boring a hole for oil, we happen to penetrate some of the lower portions of the cavity occupied by the oil (Fig. 89, b). The elastic pressure of the confined gas above will at once force the oil up, and produce a spouting or blowing well. The flow must necessarily subside by degrees as the confined gas, by the escape of the oil, acquires more space for its accommodation. It may con-

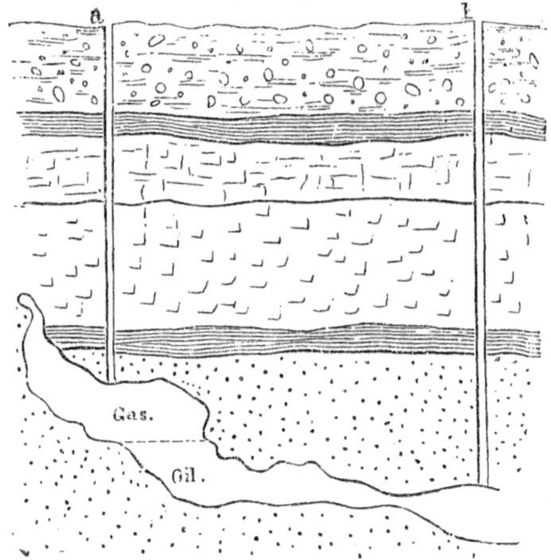

Fig. 89. Two Oil Wells.
a. A blowing well. b. A flowing well.

tinue, however, until the cavity is exhausted of its oil, after which pumping will be of no avail. If the confined gas attains its equilibrium before the oil has been completely forced from the cavity, it is evident that the remainder must be obtained by pumping. There is no cavity so large, however, as not to be destined to ultimate exhaustion. Every oil well, of whatever class, is destined to abandonment. It is true that Nature is constantly at work replenishing the exhausted reservoirs, but her accumulations are slow. Her working days are centuries.

Intermittent wells appear to act in some cases precisely after the manner of intermittent springs. More frequently, however, it is manifest that the combined action of gas and oil produces the phenomenon. In boring a well, suppose a stream of gas is struck one hundred feet from the surface of the rock, and a small stream of oil twenty feet below the gas. The entrance of oil fills twenty feet of the hole, and begins to submerge the fissure at which the gas is escaping. The gas forces its way through the oil with a sputtering sound, bubble after bubble rising to the surface. As the oil ascends, the gas makes louder and louder complaints, till finally, summoning all its accumulated energies, it hoists the superincumbent column of oil to the surface, and pours it out in a stream of a few seconds' duration. The flow then ceases, and the same operation begins to be repeated. After a minute or more of renewed grumbling and sputtering, the pent-up gas again relieves itself, and thus the work continues. The same results would ensue if oil and gas found entrance at the same fissure, or even if the gas were admitted at any distance beneath the entrance of a small supply of oil.

The amount of oil that has been ejected from certain wells is marvelous to relate. Though Western Pennsylvania has produced numerous flowing wells of wonderful capacity, there is no quarter of the world where the production has attained such prodigious dimensions as in 1862 upon Oil Creek, in the township of Enniskillen, Ontario. The first flowing well was struck there January 11, 1862, and before October not less than thirty-five wells had commenced to drain a store-house which provident Nature had occupied untold thousands of years in filling for the uses—not for the amusement—of man. There was no use for the oil at that time. The price had fallen to ten cents per barrel. The unsophisticated settlers of that wild and wooded region seemed inspired by an infatuation. Without an object save the gratification of their curiosity at the unwonted sight of a combustible fluid pouring out of the bosom of the earth, they seemed to vie with each other in plying their hastily and rudely erected "spring-poles" to work the drill that was almost sure to burst, at the depth of a hundred feet, into a prison of petroleum. Some of these wells flowed three hundred and six hundred barrels per day. Others flowed a thousand, two thousand, and three thousand barrels per day. Three flowed severally six thousand barrels per day; and the "Black & Mathewson" well flowed seven thousand five hundred barrels per day! Three years later, that oil would have brought ten dollars per barrel in gold. Now its escape was the mere pastime of full-grown boys. It floated on the water of Black Creek to the depth of six inches, and formed a film upon the surface of Lake Erie. At length the stream of oil became ignited, and the column of flame raged down the windings of the creek in a style of such fearful grandeur as to admonish the Canadian squatter of the danger, no less than the inutility and wastefulness, of his oleaginous pastimes. From detailed determinations, I have ascertained that, during the spring and summer of 1862, no less than five millions of barrels of oil floated off upon the water of Black Creek—a national fortune totally wasted, as inherited fortunes are apt to be wasted, by those not educated to an understanding of the amount of labor and time consumed in the accumulation of such fortunes. [See Appendix, Note VIII.]

The general conditions of oil-accumulation may be thus epitomized:

1. A formation containing the material for the production of oil by slow spontaneous distillation.

2. A porous formation or reservoir above the mother rock, or within it, in which the oil may be accumulated.

3. An overlying impervious formation, which shall prevent the escape of the product to the surface of the earth.

4. A dome-shaped conformation of the impervious roof, which shall prevent the lateral escape of the oil, or its dissemination through spaces too extensive.

The failure of either one of these requisites will convert all the other indications into illusory and seductive temptations. Some of these conditions it is impossible to ascertain without considerable general geological knowledge, and a minute acquaintance with the local geology of the region under consideration.

In determining upon the first condition, it is necessary to know what are the characteristics of a formation containing the materials for oil, and what particular formations in the chronological series have been proven to contain such materials. Experience has shown that comminuted vegetable matters disseminated through a matrix of fine argillaceous materials, and forming a black, or carbonaceous, or bituminous shale, are the chief source of supply in all the productive regions of the United States and Canada. The intimate mixture of vegetable and argillaceous particles seems to facilitate the chemical rearrangements involved in oil-genesis. Pure vegetable matters form beds of coal, in which the organic material is approximately fixed. The distinction between the "splintery" and "fatty" coals corresponds with their difference in purity. Cannel coals are highly carbonaceous shales. Like the "black shale" of the West, they afford copious supplies of liquid hydrocarbons.

The mother-rock of the oil in some of the most productive regions of the continent seems to be the "black shale" of the West, which is the Genesee shale of the New York geologists. This fact was first pointed out by my friend Professor Newberry, now of Columbia College. I have little doubt that this formation affords the oil obtained in Northwestern Pennsylvania; parts of Enniskillen and Bothwell, in Ontario; in Eastern and Central Eastern Ohio; in the Glasgow region of Southern Kentucky, and in Northern and Middle Tennessee. It is also probable that it supplies the oil in most of the wells of Southwestern Pennsylvania, West Virginia, Southeastern Ohio, and Northeastern

Kentucky. These wells, however, are bored through some portion of the coal-measures or underlying conglomerate, though it is not certain that any important supply has been reached by any well which terminated within the range of the coal-beds.

In the lower portion of the Hamilton group is another black shale, known in New York as the "Marcellus Shale," but assuming at the West so calcareous a character that it has not been generally distinguished from the proper limestones of the Hamilton group. It presents itself in Michigan and Ontario as a mass of black, shaly limestone or calcareous shale, overlaid by the thin-bedded and argillaceous limestones of the Hamilton group proper. I am led to regard this formation as the chief source of petroleum in the Enniskillen and Bothwell regions of Ontario.

In the Cincinnati group is another black shale which is believed to supply the wells in the Burkesville region of Southern Kentucky, and on Manitoulin Island of Lake Huron. Not unlikely, some of the impure coals of the sub-conglomerate series have afforded supplies to wells terminating in the conglomerate in West Virginia and the neighboring portion of Kentucky. The oil springs of California are supplied from formations of much more recent date.

No limestone is known to be the mother-rock of large supplies of petroleum. It is true that the Corniferous limestone is saturated and blackened in many localities by the presence of bituminous matters; and it is true that this formation lies beneath the productive oil regions of West Virginia, Western Pennsylvania, and Enniskillen, in Ontario. It is due, also, to one of the most eminent authorities in chemical geology to state that Dr. T. S. Hunt entertains the opinion that the Corniferous limestone is probably the source of petroleum in the several regions named, and especially in Ontario. He has embraced numerous opportunities to expound and enforce his views, insomuch that the conviction has obtained great currency in Canada that this limestone is the principal source of petroleum in that province. Under this conviction, scores of oil wells have been bored throughout the belt of Canadian territory immediately underlaid by the Corniferous limestone. If this formation, say they, is the source of the oil obtained at Enniskillen, where it lies five hundred feet from the surface, let us proceed to some region where this formation approaches nearer the surface, and thus save several hundred feet of boring. Though this reasoning has been put in practice in multitudes of cases, both in Canada and the United States, I am not aware of a single well bored in the Corniferous limestone that has produced sufficiently to pay expenses. I do not regard the inference acted upon as legitimately drawn from Dr. Hunt's views; for he must perceive that, even were this limestone the source of petroleum-supplies, it must have evaporated throughout the regions of surface-outcrop of the formation.

But the Corniferous limestone seems not to be the source of petroleum-supplies even in those regions where the superposition of another formation has arrested wastage. If it were the source of such marvelous quantities as have been drawn from the Canadian strata, its own cavities and interstices should certainly be charged with the liquid. To test this precise question, a " test well" was bored at Enniskillen at the joint expense of parties interested, and was continued over two hundred feet in this formation; but from the time of entering it the signs of oil were materially diminished instead of increased. The Corniferous limestone has also been penetrated at St. Clair, in Michigan, under circumstances as favorable as possible for the discovery of any great quantities of oil which may be stored up in its recesses. The salt well at that place extended through the whole thickness of the Hamilton group and the Corniferous limestone, but with nothing more than continued "signs" of oil. The same was done in the salt well at Port Austin, and in the deep boring at Jackson, in the same state.

In addition to these and other negative evidences that the view of Dr. Hunt is untenable, we have the demonstration of experiment upon the constitution of limestones and black shales. The amount of oil that can be extracted from any sample of the Corniferous limestone is utterly insignificant in comparison with the amount obtainable from the Genesee shale. It is a matter of ocular demonstration that the Genesee shale incloses a vast supply of the material for petroleum-making, while the Corniferous limestone contains almost none of the material, and comparatively little of the generated product. I shall insist, then, with my distinguished friend Dr. Newberry, late President of the American Association for the Advancement of Science, upon the correctness of that view which regards the black shales as the chief generators of supplies of native petroleum. I dissent from the position of Dr. Hunt with the utmost deference to the weight of his authority, and only because I have enjoyed opportunities seldom equaled to examine the geology of all the "oil-regions" east of the Rocky Mountains.

The petroligenous formation may be present in the absence of reservoirs for the reception of the product. There is a well-defined belt along the eastern and the western slopes of the lower peninsula of Michigan which is immediately underlaid by the Genesee shale, capped by a deposit of argillaceous surface materials adequate to prevent wastage; but no oil has accumulated, because no space has been provided for it. In some portions of the Enniskillen region, twenty miles distant, the geological conditions are perfectly identical, except that a bed of gravel lies at the bottom of the drift materials, and immediately upon the

Genesee shale. This bed of gravel is the reservoir, and becomes charged with a supply of thick petroleum called "surface oil." Some wells have yielded thousands of barrels of surface oil. It may be necessary to add that in some portions of Enniskillen the Genesee shale has been removed, and the surface wells are evidently supplied from the lower Marcellus shale, which also stocks the crevices of the Hamilton limestones. In Venango County, Pennsylvania, and Trumbull, and Knox, and contiguous counties in Ohio, the Genesee shale is overlaid by porous sandstones which serve as reservoirs of the oil. In the Glasgow region of Southern Kentucky, the formation overlying the Genesee shale is the Mountain limestone; but this is in places arenaceous, and in others vesicular and cavernous, and thus furnishes the requisite conditions of oil-accumulation. In one instance at least, in that region, the Genesee shale itself affords the reservoir for the storage of its productions. In West Virginia the oil seems to accumulate in the conglomerate at the base of the Coal-measures. The same is the case in Southwestern Pennsylvania, Southeastern Ohio, and Northeastern Kentucky. The reservoir in the Burkesville region of Southern Kentucky is found in the shattered shaly limestones of the Cincinnati group. These are reproduced in physical characters in the shattered shaly limestones of the Hamilton group, which serve as the place of deposit of the oils of Ontario.

I close this sketch of the geological phenomena of petroleum by presenting a synopsis of oil regions and the formations tributary to their supplies.

I. The black shales of the Cincinnati group afford oil which accumulates (1) in the fissured shaly limestones of the same group, and supplies (A) the Burkesville region of Southern Kentucky, and (B) Manitoulin Island in Lake Huron.

II. The Marcellus shale affords most of the petroleum

which accumulates (2) in the fissured shaly limestones of the Hamilton group, and thus supplies (C) the Ontario oil region, locally divided into (a) the Bothwell district, (b) the Oil-Springs district, and (c) the Petrolea district.

The Marcellus shale affords also a large portion of the oil which accumulates (3) in the drift gravel of the Ontario region.

III. The Genesee shale, with perhaps some contributions from the Marcellus shale, affords oil which accumulates (4) in cavities and fissures within itself in (D) some of the Glasgow region of Southern Kentucky.

It affords also the oil which accumulates in (5) the sandstones of the Portage and Chemung groups in (E) Northwestern Pennsylvania and contiguous parts of Ohio.

It affords also the oil which accumulates in (6) the sandstones of the Waverly (Marshall) group, in (F) Central Ohio.

It affords also that which accumulates in (7) the mountain limestone of the Glasgow region of Kentucky and contiguous parts of Tennessee, as also some of that which is found in the drift gravel of the Ontario region.

IV. The shaly coals of the false Coal-measures, aided, perhaps, by the Genesee and Marcellus shales, seem to afford the oil which assembles in (8) the coal conglomerate as worked in (G) Southwestern Pennsylvania, (H) West Virginia, (I) Southern Ohio, and the contiguous but comparatively barren region of Paint Creek, in Kentucky.

V. The Coal-measures may perhaps be regarded as affording a questionable amount of oil, which may have been found within the limits of (9) the Coal-measures in the West Virginia and neighboring regions.

From this exhibit it appears that the principal supplies of petroleum east of the Rocky Mountains have been generated in four different formations, accumulated in nine different formations, and worked in nine different districts.

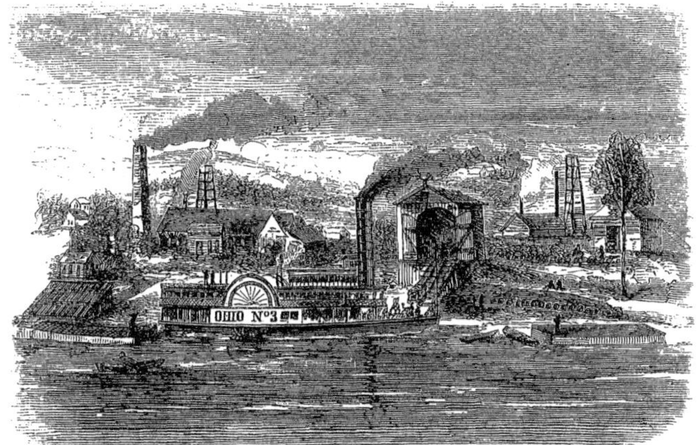

Fig. 90. View of the Salt Works at Mason City, West Virginia.

CHAPTER XXVI.

SOMETHING ABOUT ROCK-SALT AND GYPSUM.

COMMON salt, upon which the chemist has imposed the more dignified title of chloride of calcium, is a mineral almost universally distributed through the stratified portion of the earth's crust. Like those other substances of universal utility to man—petroleum, coal, iron, water, and lime—it is supplied by Nature to every habitable region of the terrestrial surface. Like lime, which is the chief constituent of the bones and teeth of man and the other vertebrates, the shells of molluscous animals, and the mountains of coral accumulations reared in the bottom of the sea, common salt also subserves the necessities not only of man, but of the quadrupeds and various other terrestrial animals, including insects, and is the characteristic constituent of sea-water, the home of two thirds of all the animals now existing, and a much larger proportion of the animals of former geological ages.

The salt of the rocks is the residuum of the once universal ocean. The reader will remember that reference has already been made to the origin of salt lakes, like those of Utah and the Caspian and Aral Seas. Such lakes are but remnants of the last oceanic inundation. They occupy depressions in the terrestrial surface from which there is no outlet. If, like Lake Superior, they had been drained to the sea, the original saline waters would long since have been replaced by fresh waters from the clouds.

In consequence of the changed condition of the earth, the amount of evaporation from the surfaces of these inland seas has generally exceeded the contributions of fresh water from the clouds. Their saltness has therefore been intensified, and, in many cases, a deposit of crystallized salt has been formed upon the bottom and around the shores. Indeed, there have been salt lakes that are now extinct, in consequence of the exhalation of their waters; and in the place of each remains a salt plain, the surface of which is composed of salt and the other mineral constituents of the ancient sea-water, variously intermingled with argillaceous matter washed in from the surrounding country. The interior of the American continent furnishes abundant phenomena of this kind, stretching from Utah, through the Great American Desert, to Mexico. Such products are the residua of salt lakes which have evaporated since the surface received its existing configuration. It will be remembered, however, that extensive salt-beds exist in Nevada, which are derived from the leachings of the saliferous strata of the mountains; and it may be that some of the ancient salt lakes of the region were supplied with salt to some extent by contributions from similar sources. This, nevertheless, would not prove that all salt lakes have been similarly fed. Besides, if it should appear that they are, we have still to account for the existence of strata of salt packed away among the solid rocks; and there is no explanation so natural and so consonant with what we know of the history of the world as the doctrine of evaporated sea-waters. How the waters of the sea came into possession of their saltness is a question of primeval chemistry to which allusion has heretofore been made. It was the resultant of the chemical actions which took place between the fire-born rocks and the atmospheric acids washed down by the primeval rains, and gathered with "the gathering together of the waters."

Salt lakes, or detached outliers of the great ocean, have existed in all ages since the continents began to shed the ocean's waters from their backs. In the age just preceding the last, an inland sea occupied the region of the upper waters of the Missouri River; and, a little earlier, the same sea extended a few hundred miles farther south, over the country of the "Bad Lands" of Dakotah. In the middle ages of the world's history, the evaporation of salt lakes or bays more or less shut off from the ocean, and the bedding of their saline constituents, was a phenomenon of so frequent occurrence as to constitute the most prominent feature of an entire group of strata. This group has consequently been styled the "Saliferous system." The saliferous beds of this group are extensively worked for rock-salt over a territory stretching along both sides of the Carpathians, embracing the mines of Wallachia, Transylvania, Galicia, Upper Hungary, Upper Austria, Styria, Salzberg, and the Tyrol. In England they are mined in the counties of Cheshire and Worcestershire. In the United States we find saliferous beds of the same age extensively distributed over the region between the Mississippi River and the Rocky Mountains.

Descending in the series of American strata, we find the Coal-measures in certain regions—or rather the conglomerate at the bottom of the Coal-measures—to be copiously saturated with brine; and in the lower peninsula of Michigan, the Marshall sandstones at the bottom of the Carboniferous system are a reservoir of saline accumulations. Still lower, the American geologist finds the Salina group of the Upper Silurian system the source of supplies of brine throughout a wide extent of territory.

The attempt has been made to explain the existence of saliferous and gypsiferous deposits by reference to chemical reactions transpiring subsequently to the solidification and upheaval of the strata; but I am led to regard the presence of sulphuric acid and other chemical constituents of gypseous and saliferous formations as products of the decomposition of previously existing gypsum and salt, rather than the agents employed in the present generation of them.

The body of water in which the saliferous materials accumulated may have been a bay or sea having imperfect communication with the ocean. Under ordinary circumstances, the evaporation from the surface of the bay would exceed the supply from atmospheric sources, and there must arise, consequently, a gradual influx of sea-water from the ocean. The bay-water would finally reach such a state of condensation as to begin to precipitate its least soluble constituents. These would be mingled with the ordinary sediments and *débris* of saline waters. This process continuing, the condensation would reach, in succession, those stages at which peroxyd of iron, gypsum, common salt, and Epsom salts would be crystallized and deposited around the shores and bottom of the bay, and mingled with the argillaceous mud brought in by the influx of surface waters. These substances are all constituents of sea-water. In course of time, the bottom of the bay may have been converted into dry land through the course of continental elevation. On the other hand, it may have been subjected to a depression of such an extent that the region became again the site of the open sea; and sediments of later date were accumulated upon the top of strata inclosing rock-salt and gypsum.

The preservation of the saline constituents of a formation thus originated must be conditioned on the vicissitudes to which it was subsequently subjected. It is obvious that the original conformation of the saliferous strata must have been somewhat dish-like or depressed in the centre, with the borders elevated. In the uplift of the continent, all portions may have been simultaneously raised, or the formation may have become decidedly tilted. In the filtration of surface waters through the interstices of the strata, it is obvious that any formation so posited as to permit a flow of water through it, either vertically or laterally, must have all its soluble constituents dissolved out. A vertical leaching may simply transfer these constituents to some lower formation underlaid by an impervious floor. A lateral drain may discharge the soluble contents at the surface of the earth, and thus, by degrees, restore them to the ocean, their ancient home. Hence many strata now destitute of either salt or gypsum may have embraced both at the time of their origin. In others we witness these substances—especially the gypsum—in process of disappearance.

In case the gypseo-saliferous formation has retained its centrally depressed conformation (compare Fig. 91), it is apparent that the saline constituents held must be unable to escape by drainage. Surface waters will fall upon the belt of outcrop of the formation, and may find their way to the interior in sufficient quantity to redissolve the soluble matters. This having been done, however, the saturated solution will charge the interstices of the formation, and

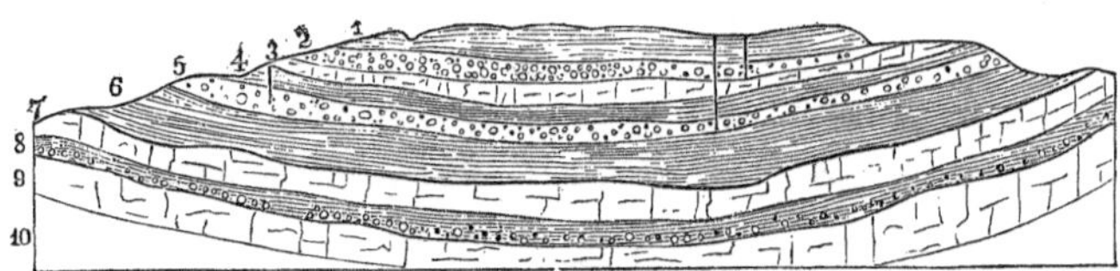

Fig. 91. Section from East to West across the lower peninsula of Michigan. 1, 2, 3, 4, 5, 6, etc. The several groups of strata from the Coal-measures to the Lower Silurian.

will suffer the diluting influence of surface waters only around the outcropping borders. Fresh water will float as a distinct stratum upon a stratum of strong brine. The deepest parts of a saliferous formation must consequently contain the strongest brine. The place of salt springs will naturally be along the outcropping belt of the formation. They are the mere overflow of the basin caused by surface rains. The region over the most depressed portion of the basin, and consequently over the deposit of strongest brine, is likely to be completely destitute of salt springs. The position of the brine-supply is therefore a problem for strictly geological determination. It is an induction from the general geology of the entire region. Superficial investigators have frequently instituted borings in the vicinity of brine springs. Inevitably such explorations must immediately pass below the source of brine-supply, and must prove unsuccessful, unless they can be extended to some more deeply seated basin, whose outcropping rim is comparatively remote. The most successful salt wells are those which are bored far from surface indications, in places pointed out by geology as located over the central portion of a saliferous basin.

From the conditions of the case, it is almost a hydrostatical impossibility that a good brine well should be a flowing well. The strong brine must be pumped up from the bottom. It may be asked why, if the borders of the basin be elevated, will not the brine be forced up by hydrostatic pressure? I admit that if the borders were elevated *on all sides* above the place of boring, such would be the case. But if the borders were thus elevated, we should have an area without surface drainage; and, instead of being a place for salt-making operations, it would be the bed of a sea or lake. The supposed condition is therefore incompatible with the hypothesis of well-boring. If we assume the existence of a single gap in the encircling rim through which the surface waters may be carried off, it must be borne in mind that this gap will also drain the brine-formation to the same level. The sheet of brine will not, therefore, rise to a higher level than the place of boring; and if the elevated rim become charged with fresh waters, they can be of no avail for hydrostatic pressure, since the notch is an outlet through which the pressure would find relief at that level. Of necessity, then, the place of boring must be somewhat higher than the continuous rim of the saliferous basin, and the brine can only be brought to the surface by the pump. In penetrating to the deep-seated reservoir of brine, other water-bearing strata may be passed whose elevation, at some point more or less remote, may be such as to originate an Artesian overflow. In working the deep brine, this water must either be stopped off, or a closed tube must be sunk through the midst of it to the brine formation, where it must be closely packed around, to prevent communication with the fresh waters above.

One other consideration should be mentioned. The brine is not always—nor generally—found in the formation in which the salt was originally deposited. When, on the elevation of the continent, meteoric waters percolated through the strata and redissolved the salt, the solution would be retained in the same formation only on the condition that it was underlaid by an impervious floor. This is generally the case with the soluble matters of the Salina group. If, however, the saliferous formation were underlaid by a porous sandstone, this would become the reservoir in which the leachings of the saliferous formation would be preserved. Thus the Conglomerate becomes in Ohio and Michigan the reservoir for the Coal-measures (Fig. 91). Borings for salt must necessarily extend to the formation in which the brine is accumulated. This is commonly designated the salt-rock; but it is not necessarily the mother-rock of the brine.

Such I believe to be a true account of the natural history of rock-salt and native brines. The phenomena of gypseo-saliferous formations seem incompatible with any other explanation. 1. The rocks composing these formations are regularly stratified, and furnish the usual indications of sedimentary origin. The beds of gypsum and of rock-salt, when existing, are entirely conformable with the argillaceous strata, and approximately coextensive with them. On this theory, having ascertained the existence of a brine formation on the west side of the State of Michigan, I successfully predicted its discovery on the east side. The extensive gypsum beds, also, of the east side were brought to light by a similar prediction based on the same theory; and I have evidence that the gypsum formation of Grand Rapids and Alabaster, on opposite sides of the state, is absolutely continuous beneath all the intervening region. 2. Gypseo-saliferous formations contain all the well-known constituents of sea-water.' I do not consider it likely that these constituents would be associated in the same way in both cases, unless the one were the historical consequent of the other. 3. The order of arrangement of these constituents is the order of their solubility. When natural brines are operated upon for salt, the least soluble constituent, peroxyd of iron, first precipitates. This is separated in the tanks before the brine is introduced into the kettles. Next, after the boiling begins, the gypsum is deposited, forming a crust upon the inside of the kettle. Next in order, common salt begins to fall down. After most of this has been crystallized out, there still remain chloride of calcium and sulphate of magnesia (Epsom salts), constituting the "bitterns" of the salt manufacturer. Further evaporation would separate the Epsom salts next in order. These several substances are arranged in the same order in natural brine-formations. At the bottom we find red clays, colored, of course, by a deposite of peroxyd of iron. Next above are clays containing gypsum. In many instances the sea-water was so clear that the gypsum was deposited in pure crystallized beds, from ten to thirty feet in thickness. Above the gypsum, in formations that have not been leached by surface waters, we find the great mass of rock-salt. Still higher are shales and limestones, containing impressions, at least, of the needle-shaped crystals of Epsom salts which were once there, but have been dissolved out by the waters which have since saturated the strata. 4. The very discontinuity of the gypsum beds in certain formations, as the Salina group in New York, is accompanied by such phenomena as to prove that the gypsum was once continuous, and is being gradually dissolved out. The overlying and underlying clayey beds assume the place of the dissolved portions of the gypsum. The remaining lenticular masses of gypsum become thus inclosed by tortuous layers of clay and shale, which look as if they had been primarily deposited about these masses, and adjusted to them. If the overlying clay be most yielding, the vacated space is mostly filled by an inflection from above. If the underlying clay be most yielding, the inflection is from below. Thus abrupt loops of clay or shale rise or sink into the spaces between isolated gypsum-lenticules. 5. Gypseo-saliferous formations are generally of local extent in one direction or in both, indicating that they were accumulated in a restricted portion of the ocean.

The productive salt formations of the United States are three. The Salina group is the source of supply of brine and gypsum to Onondaga and Cayuga Counties, New York. The vast manufacture of the Empire State is based upon this supply. Only the northern rim of the basin or formation is known (Fig. 92). Its outcropping edge was deeply excavated by the agencies of the ice-period, and the excava-

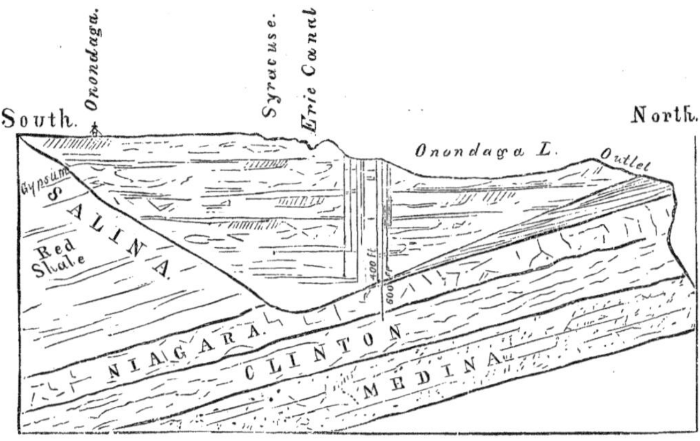

Fig. 92. Longitudinal section of the Onondaga Salt Basin (from Superintendent's Report for 1857), showing the ancient excavation of the outcrop of the Salina group, now filled with gravel and clay, and saturated by an exudation of brine from the old stump of the formation.

tion was filled with gravel. The overflow from the notched rim of the basin saturates the gravel, and thus forms a vast inland salt-marsh. The strongest brine settles to the bottom of this basin, and is reached by wells of the ordinary kind, and pumped out. It seems inevitable that a supply obtained under such geological circumstances must be liable to rapid exhaustion. The facts show that the strength of the brine is gradually diminishing. The same formation affords brine and gypsum in the vicinity of the Grand River of Ontario, and rock-salt and strong brine at Goderich. It is worked for gypsum in the vicinity of Sandusky, Ohio. It underlies the whole of the lower peninsula of Michigan (Fig. 91) in the form of a vast basin, whose borders come to the surface at Milwaukee on the west, Mackinac on the north, the Grand River of Ontario on the east, and Sandusky on the southeast. This great salt basin has been penetrated, under the guidance of geology, at St. Clair and at Point aux Barques, and successful wells eleven hundred feet deep are now in operation. A new well is about to be put in operation at Mount Clemens, in Macomb County, and others are in progress at various points.

The next saliferous formation, in ascending order, is one which is peculiar to the lower peninsula of Michigan, and has hence been styled the "Michigan Salt Group." Its geological position is between the Marshall sandstone and the Mountain limestone. It underlies, like a great dish, nearly the whole of the peninsula. Its outcropping rim is marked by a circuit of salt springs. Filtration has leached out most of the brine into the underlying sandstones. The gypsum, however, mostly remains in the formation, and is extensively worked. The wells of East Saginaw and Saginaw City are supplied from this formation. As in the case of the Salina basin, this one is reached by deep borings over the most depressed portion (Fig. 91). These borings were originally undertaken as the result of a pure geological induction, and strong and copious brine was obtained at the depth of about eight hundred feet. The first rock, even, was one hundred feet from the surface, and the whole thickness of the Coal-measures had to be crossed. I consider such successes ample vindication of the utility of geological science. The geological survey, of which this was but one of the results, cost the state five thousand dollars. The discovery of brine in the Saginaw Valley has added two millions of dollars to the capital of the state.

The next conspicuous salt formation in ascending order is the Coal-measures. The reader who recalls the surface conditions under which the coal was formed will at once perceive that there must have been a great concentration of sea-water in the remote and somewhat isolated lagoons and marshes in which much of the materials of the coal formation were accumulated. It will be noticed, also, that the associated strata are here, as elsewhere, predominantly argillaceous. As the Coal-measures are universally underlaid by the great Conglomerate, this becomes the reservoir in which the saline solutions from the Coal-measures accumulate. The Conglomerate is the "salt-rock" of Ohio, West Virginia, and Northeastern Kentucky. It also underlies a large central area in the peninsula of Michigan, and thus constitutes the third great salt basin within the limits of that state, each underlying the same central area. The shallow wells at Bay City, Portsmouth, and the Lower Saginaw River generally, are supplied from the Conglomerate. The deeper ones at the same places are supplied from the next basin below. The gypsum is generally dissolved out of the Coal-measures, but in Western Iowa it still exists in vast quantities.

In Southern Kentucky, and Northern and Central Tennessee, brine is obtained by boring into the "Silicious group"—a local name for certain members of the Mountain limestone. I will not attempt to decide whether this brine proceeds from the Coal-measures or the False Coal-measures, or has had an independent origin.

In Texas, Colorado, and Kansas, salt and gypsum are supplied in vast quantities from formations of Mesozoic age, as in Europe. In some of the Gulf States, especially Alabama, tolerably strong brine is obtained by boring into the lower argillaceous and arenaceous strata of the Cretaceous system. On the island of Petite Anse, on the coast of Louisiana, nine miles south of New Iberia, is a remarkable deposite of rock-salt, till very recently the only one known to exist east of the Rocky Mountains. Underneath the soil of at least one hundred and forty-four acres of this island lies a solid bed of pure rock-salt, in which pits have been sunk to the depth of thirty-eight feet without reaching the bottom. The mass of this salt is below high water. It is overlaid by about nineteen feet of clay, gravel, sand, and surface soil. Not less than twenty-two million pounds of salt were removed from the island during the eleven months previous to April, 1863. The supply is probably inexhaustible. This extraordinary mass may occupy the site of an ancient bayou, the bottom of which has been elevated, while the contiguous shores have been either eroded or depressed, so that the land and water have exchanged places. It is the opinion of Dr. Goessman, however, that it is "a secondary deposite, resulting from the evaporations of brine-springs originating from beds of rock-salt in some older geological formation, and not a direct residuum from any sea."*

* Report of the American Bureau of Mines, 1867. Professor E. W. Hilgard has also made an examination of this deposite. (See *Amer. Jour. Sci. and Arts*, Jan., 1869, p. 77.)

CHAPTER XXVII.

METHOD IN THE GROWTH OF CONTINENTS.

HOW impressive the unity of purpose with which Nature has pushed forward the consummation of her vast schemes! Ends have been foreshadowed through almost an eternity of years, while the all-directing Mind has steadily controlled the ministering forces, in the midst of millions of disturbing agencies, till the premeditated work has been accomplished. We witness in the plans of the Infinite Architect the same intelligent cohesion of parts as in a well-laid human scheme; and while the relations of certain events far transcend the scope of our reason, and the perfection of contrivance is immeasurably superior to that of human designs, we understand enough and measure enough to know that a philosophy which is at once human in its method and divine in its comprehension underlies the whole chain of natural events. There is a logical relationship of things established by God and recognizable by man, and the sequences of events are ofttimes so clear that even finite intelligence is able to penetrate the future and unveil plans existing only in the Infinite conception.

This ideal connection of the parts of the Creator's universe is, perhaps, best traced among organized beings, but I propose first to point out its existence in the history of inorganic nature. The infinitely diversified features of the earth's surface have been wrought out by the operation of a few principles working through ages in definite modes. We see that certain rocks bear the evidences of their sedimentary origin. We look about, and find sedimentary accumulations still forming and hardening. We look back, and ascertain that the same processes, continued through ages of the past, have piled up thousands of feet of rocky beds, in which still slumber the mummied forms of the primeval world. We see that certain rocks bear the marks of fire. We plunge our hands into a thermal spring, and gather intimations of internal heat. The molten eructations of a volcano demonstrate the continued existence of melted rocks. If masses of igneous origin have cooled from a state of fusion, who can say that they have not cooled from that higher temperature at which we know that rocks and all other things can subsist only as vapor? Do we find rocks existing in that condition? Yes; worlds still exist as igneous vapors. Here, then, we may assume our starting-point. A world of airy flame, after ages of cooling, gathered a liquid nucleus at its core—a globe of molten rock, wrapped in a glowing atmosphere of all that remained as vapor. Next, a fiery floor congeals over the surface of the burning tide; the burning tide, as if in rage, lashes it to fragments, and the abated heat allows them to be recemented. When the hotter fires had been quite imprisoned in the strengthening crust, dews began to gather in the upper air, and streaks of haze barred out the burning beams of the lurid sun. Rains fell upon the fervid crust, to waste themselves in sudden vapor, and return to the attack upon the crust. Gleams of electricity lighted the misty drapery of this geologic night, while the thunders of Nature's ordnance echoed through the caverns of the clouds.

A rain of acid waters at length got the mastery of the wrinkled surface, and every ravine and valley witnessed the race of the rivers for the lowest levels. Every watercourse bore onward its freight of sediment, the materials of the masonry of continents. The filmy ocean swallowed the rivulet, crawled over the hill-top, and embraced the world. The world, in turn, opened its wide and rocky jaws and swallowed the ocean—and another ocean laved the face of Nature.

In the progress of events, an occasional ridge of barren granite lifted its back permanently above the level of the sea. As the liquid core contracted, the surplusage of the enveloping crust was absorbed by the wrinkles already existing, and thus the granite backs rose higher and higher. As the ridges were higher raised, and the valleys deeper sunken, the accumulated oceans pressed heavier and heavier against the slopes of the rocky beds, and the gathered sediments of ages weighted the ocean's floor with a burden which easily outweighed the crust which bridged the hills. And thus it was that the valleys were ever deeper sunken, and that which was at first an insignificant wrinkle became at last a stable mountain. From the coast of Labrador southwest along the Laurentian Hills we tread upon that ancient summit which was the first-born of Old Ocean. From the far northwest it comes down to us with the same time-worn record written on its weathered brow, while a chain of noble lakes fringes the angulated ridge along its western branch, and the eastern bathes its feet in the waters of the St. Lawrence. As the flowers of one spring-time foretell the forms which will reappear when spring-time comes again, so this ancient germinal ridge was but the first blooming of a continent; and when the circle of a geologic year was run, the rocky leaves of the growing continent unfolded themselves again in their appointed fashion. Note the parallelism of that primeval ridge with the present shores of the Atlantic and Pacific. When we know that each successive revolution of the globe has but rolled the waters of the oceans farther to the southeast and southwest, do we not perceive that the deep ocean's bed has ever been the deep ocean's bed, and that the first ridge of land was the nucleus of the continent, and the trend of its shores a prophecy of the coast-lines of our day?

Here, then, immeasurable ages before the creation of man—before even a living thing had crawled in the waters of the sea—Nature had distinctly staked out the birthplace of American freedom, and fenced in one inclosure the vast area between the Atlantic and the Pacific—between the great lakes and the Mexican Gulf—and forebore to raise a single separating barrier from one extreme of the empire of freedom to the other. And, through all the chances of following revolutions, she has never erected an Alpine boundary to thwart her purpose in the unity of the continent.

By successive upheavals belt after belt was added to the area of the land. Even a phase of continental history which seems somewhat exceptional was wrought out by the strictest adherence to the established methods. When the time arrived for the creation of land animals, the shrinkage of the nucleus had proceeded to a point which subjected the crust to the most enormous lateral pressures. Uneasy in every attitude, it maintained a perpetual oscillation—I say perpetual, though in movements so vast a hundred years are as a moment. Vegetation, which was appointed the scavenger of the atmosphere, gathered up its freight of carbon, and a well-timed subsidence of the surface inundated the carbonaceous accumulation, and buried it in mud and sand far from the reach of the destroying influence of the atmosphere. A hundred times the process was repeated; and so it happened that when the atmosphere was purified, the tension of the crust could be no longer borne, and one grand convulsion rolled up the Appalachians in their hundred folds; and there, nicely assorted between the rocky leaves of the mountains, were the layers of carbon, changed from the poison to the comfort of the coming man!

To recount the events of the following ages is to repeat the story of the past. By-and-by the plastic hand of Nature had moulded the continent to its destined features. It seemed to need but man to be a finished work. But the Creative Architect contemplated a higher finish than human wisdom could have contrived. Now that the Atlantic and Pacific had completed those portions of the continent in their more immediate vicinage, it remained for the smaller sea which surrounds the pole to develop by its pressures the northern slope of the land, and thus to become the remote agent in strewing the surface of the rocks with an arable soil. The uplift of the arctic regions brought on the reign of ice, and wintry devastation swept over the late verdant landscapes. The downthrow of the Arctic highlands ameliorated the climate, and Spring again visited the icy fields. The movements of ice and water left the surface covered with cubic miles of rubbish produced from the destruction of the underlying rocks. But the entire continent was destined to a new baptism. The once forbidden ocean was readmitted to career in triumph over states that had long ago been reclaimed from his dominion. Michigan disappeared beneath the wave, and Ohio, and Pennsylvania, and New York, and Canada. The entire northern and middle regions of the continent sank down to a level lower than had been reached since the deposition of the coal. Then, in due time, began the last resurgence of the land. By degrees the finny waters shrunk back nearly to their former lines. Now the river channels were dug out; and now the Niagara began anew to plow its stupendous gorge. Unknown ages passed, and man assumed the sceptre of the earth.

Fig. 93. Professor James D. Dana.

With what fidelity has geology deciphered the records of this wonderful history! We marvel that so many secrets of the silent ages have been found out. And yet we run over their chronicles as if but the annals of the last year. How immense a field for the imagination to sweep over! What amazing intervals of time to contemplate! what gigantic operations to trace! And yet we behold from the beginning the action of the same physical forces as are in action to-day. The immutable and omnipresent forces of chemistry first held the elements under sway. Affinity, gravitation, caloric, electricity, in their varying operations, have wrought out the diverse phases of the modern earth. The plan of operations has been equally uniform. Igneous forces pressing upward—oceanic waters bearing downward and outward. An incipient wrinkle, a growing ridge, an upheaved cordillera. The ocean bed was made for the primeval waters. The place for the continents was marked out in earliest time, and each successive event contributed consistently to the final consummation. Even their outlines were foreshadowed in the trend of those primal ridges which made a mockery of dry land before a living thing had appeared upon the earth. And when the finishing touch was to pass over the globe, we find it effected by the same general agency as piled up miles of strata and raised granite summits to the clouds. An upheaval, a submergence, and another upheaval constitute the last three chapters of the history. Who can contemplate this identity of agencies, this persistence of plan and perfection of results, without being impressed that One Intelligence has planned the scheme and guided the blind forces from the beginning to the accomplishment of the long-anticipated end?

O

CHAPTER XXVIII.

METHOD IN THE HISTORY OF LIFE.

NATURE has always issued her bulletins. It is a most interesting fact in the history of the animal creation that Nature advertised her plans in the very earliest creative acts. In our study of the relics of the primeval ages we do not find the grand and fundamental purposes of Infinite Wisdom unfolding themselves by degrees as type after type of organic life made its advent upon our planet. It is quite true that the full development of Nature's schemes can only be apprehended in the ultimate results, and that, with our highest wisdom, we are continually surprised at the wealth of resources exposed in the unfolding of a simple plan. But Nature had her plans, and these were mature in the very beginning. All possible contingencies being foreseen, no amendments or modifications have been necessitated by the growth of successive populations and the march of human improvement. The outlines of Nature's grand methods were announced in her initial creative efforts. It was thus in the plan of continental development; it was thus in the plan of the animal creation. It is only in the infinite flexibility of her plans, and in the inexhaustible richness of their filling up, that Nature transcends all the possibilities of human expectation.

To the geologist no fact is more familiar or more patent than the simultaneous introduction upon the earth of three of the four fundamental plans of animal structure which in the following ages were to sport into the infinite variety of individual forms that diversify the surface of the earth at the present day. Saying nothing about the solitary *Eozoön*, which stands inscrutable, isolated, and mysterious in the remote ages of Eozoic Time, like a desolate islet in the midst of a dark, and trackless, and tempest-beaten sea, we find that upon the very threshold of Paleozoic Time representatives of Radiates, Molluscs, and Articulates burst into multifarious being almost simultaneously. So nearly simultaneous was the appearance of each of these types, that all hypothesis of their genealogical succession is rationally precluded. The doctrine of development finds great discountenance in the very first of the facts from which such a doctrine ought to derive its support. Later in the history of the world Vertebrates made their advent, and thus were laid the four corner-stones on which Nature has built the superstructure of the animal creation. Among all the multitudes of organic forms which have been disentombed from the cemeteries of the solid rocks, we have found none which were not conformed to one of the four fundamental types announced in the beginning. Here is no caprice, here is no chance, but the constancy, and order, and persistence of intelligence, foresight, and fixed purpose.

When this grand procession of organic forms was marshaling for its movement through time, the Supreme Intelligence sent it forward in four columns, in each of which was dominant one of the four ideas of structure. But as Nature did not range her four columns in linear order, but set them abreast of each other, so she was equally far from bringing forward the subordinate divisions of each column or plan in any thing like a fixed progressive succession. Neither the highest and most exalted forms, nor the lowest and most humble, were ordained to take absolute precedence. In the sub-kingdom of Radiates the type was introduced by Echinoderms, Acalephs, and Protozoans, the two highest and the lowest of the four classes. True coral animals perhaps made their appearance a little later. In the sub-kingdom of Molluscs all the classes stand abreast on their first advent; in that of Articulates, the two lower classes, Crustaceans and Worms, preceded by a long interval the Insecteans; and in the sub-kingdom of Vertebrates the classes followed each other in regular gradational succession. Thus we see that, so far as class-groups are concerned, it is impossible to reduce the order of succession to any general formula. How is it with the orders of the respective classes? Among Echinoderms, Cystideans appeared before the successively higher Crinideans, Starfishes, and Sea-urchins; among Acalephs, the horny Graptolites appeared before the Coral-builders; among Protozoans, the Sponges, which ally themselves to Polypi, appeared before the lowest types—always disregarding the mysterious *Eozoön*. On the whole, the order of succession among the groups, based upon relative rank, is, with Radiates, from below upward. With Molluscs we find the straight and simple Orthoceratites preceding the higher Cephalapods; the arcuate and the entire-mouthed Gasteropods leading the higher spiral and flesh-eating families; the asiphonal Lamellibranchs antedating those with more complete respiratory apparatus, and the horny-shelled *Lingula* and *Discina*, among Brachiopods, appearing before the stony-shelled and stony-armed Spirifers and Terebratulas. Among the Articulate and Vertebrate classes the gradational succession of the various orders is tolerably perfect. But I must refrain from alluding to specific facts. The following grand generalization rests on a broad survey of data upon which it would be inappropriate, in this place, to enter.

There is no successional relation between the four subkingdoms of animals, nor even between the several classes of the invertebrate sub-kingdoms; but among the orders of the several classes and the classes of the Vertebrates we find generally a progress from lower to higher in the order of introduction.

But there is another principle, complementary to this, which needs to be united to it in order to present us with a true view of Nature's method. There has generally been a downward as well as an upward unfolding of each type from the central forms in which it was first embodied. Trilobites, the first representatives of the Crustacean type, belong indeed to the lowest group, but do not lie at the bottom of the group—the lower members, as well as the higher groups, coming into being at subsequent periods. The earliest reptiles were not the lowest of the Amphibians, but Labyrinthodonts, the highest Amphibians; and from this starting-point the reptilian type expanded both upward and downward. Vertebrates began, not with the lowest fishes, but with a grade of fishes above the mean level of the type in the possession of several reptilian characteristics. From here the type rose still higher to the strongly sauroid forms, and descended to the Teliosts, or typical fishes, with their aberrant and degraded forms—the lamprey and the lancelet. We shall arrive, therefore, at the truest expression of the plan of Nature in reference to the succession of organic beings by saying that each type was first introduced at a nodal point, from which the stream of development proceeded in both directions—the lowest forms in many instances being reached only in the modern age; so that, in some cases, after the culmination of a type, it has suffered a degeneration into the lower grades already passed.

Another fact strikes us in a review of the succession of life in past time. Life has presented itself not so much in a series of sharply-restricted organic *forms*, rising or descending in regular order, as in a succession of *dominant ideas*, each in its own age expressing itself in more than one organic type. Thus, in the reign of reptiles, the reptilian idea was dominant, and we find it invading the structure of the contemporaneous fishes. Afterward the avian or ornithic idea became dominant, and reptiles were endowed with wings, and even with feathers—if we may credit the reptilian character of the Archæopteryx of Solenhofen. Still later, the mammalian idea became dominant, and the forms of the *Ichthyosaurus* and *Plesiosaurus*, but more especially of the terrestrial Deinosaurs, indexed the impress of that idea upon the reptilian class. Even in the Age of Molluscs, the dominant idea was expressed in the bivalve nature of the Ostracoid Crustaceans.

The forms styled "synthetic" or "comprehensive" types may perhaps be generalized under the formula of *dominant ideas.* Comprehensive types are those in which certain characteristics of a group are ingrafted upon a distinct though kindred stock. The Ganoid fishes are of this kind, since they combine reptilian with fish-like features. The Labyrinthodonts were comprehensive types, because they were Amphibians with the scaly covering of Reptiles. The Lepidodendra of the Coal era combined the characteristics of the Cryptogams with the foliage and general habits of the Conifers. Such a synthesis of types seems to be occasioned by the overlapping of consecutive ideas in time—a penumbra occurring while the last dominant idea is passing under the shadow of the coming one. The Pterosaurs, or flying reptiles, were the most marvelous of all comprehensive or penumbral types. On the basis structure of a reptile we find ingrafted the head and neck of a bird, the trunk and tail of a quadruped, and the leathery wings of a bat; while, not improbably, their feet were furnished with a web; so that these creatures were fitted for all elements, and showed in their structure a synthesis of features belonging to each of the four classes of vertebrates.

The dawn of a new dominant idea in Nature is often foreshadowed by penumbral types of an anticipatory character. The earliest representatives of a comprehensive type were generally prophetic. Before reptiles were created, hints of coming reptiles were dropped in the constitution of the fishes, as in the concavo-convex vertebræ of the Ganoids. The winged bird was foreshadowed by the flying and feathered reptiles; and mammals were heralded by the whale-like and paddle-bearing Ichthyosaurs. What but prophetic types were all the first-formed creatures belonging to the four grand categories of structure, from which have been developed the diversified beings of after ages? All possibilities of vertebrate existence were folded up in the constitution of the first fish which Omnipotence called into being. In the organization of those primordial Trilobites which figure in the vignette of animal history were wrapped up in potentiality all the species which creative power has since evolved from the articulate type—lobsters, barnacles, centipedes, spiders, butterflies, beetles. These forms were all in full view of the Intelligence which executed the plan that involved them, and which, in its destined unfolding, must set them free upon the earth.

Most impressive are the facts which show the ideas of the far-off coming ages wandering in advance of their time among the creations of an existing world, like streaks of morning light which herald the approaching sun through all the sky, while the world still sleeps under the reign of darkness. It is as if the thoughts of the Creator were busied with the plans of the distant future, while his hands are occupied with the work of to-day. Thus were incorporated in the organisms of one age hints of the features which were to blossom and unfold in the dominant ideas of the following one. Thus grew into being those "prophetic types" which show that *One Intelligence* has ordered creation—an intelligence to which the past and the future are both present. Here are relations of thought which proclaim in the ears of all men that chance has never swayed the sceptre of the world, and that unthinking and blind material force has only been the servant of an Intelligent Will.

Hardly less interesting are the phenomena of retrospective types. These lie on the vanishing side of the eclipse. They are the last shadows cast by a type whose central passage was ages ago. The "garpike" or "billfish" of our Western waters is a notable example of retrospective types. Some geological cycles since, the garpikes were the monarch-occupants of the waters of the earth. Helmeted and mailed in impenetrable armor, they were secure from the attacks of the most formidable foes. With jaws armed with triple rows of sharp and conical teeth, and endowed with the power of darting like an arrow through the water, there was no contemporary too swift to capture or too powerful to destroy. The meridian of this dynasty was in the Mesozoic ages. From that time its power has continued to wane; and in the present age, so far as shown by the published records of science, only the *Polypterus* of the Nile and Senegal Rivers, and the *Lepidosteus* of North American waters, survive to represent the prestige, and glory, and prowess of a reign which was once inexorable and universal. We may look upon the "billfish" of our rivers and lakes with a veneration infinitely more exalted than any belonging to the survivors of the decaying dynasties of human history. Here are the relics of empires in which the Almighty Will has wrought its own purposes; on the other hand are the ruins of fabrics built and defended at the cost of human liberty and human blood, in which human license for a time has been suffered to wrestle against the Almighty Will.

Equally profound is the lesson taught by the *Pentacrinus* of the Caribbean Sea, for it stands there the sole survivor of the Crinoids of the Paleozoic world. A delicate stony stem, affixed to the submarine soil, bears upon its summit a symmetrical cup or body, around the margin of which are supported the five stony arms which ramify into scores of fingers. The whole structure is composed of many thousands of little stony pieces, many of them handsomely sculptured, and all fitted together with mathematical precision. Dr. Buckland demonstrated that the number of separate pieces in a fossil *Pentacrinus* was more than 150,000, while M. de Koninck calculated that an adult specimen of the same species (*Pentacrinus Briareus*) was composed of not less than 615,000 separate pieces. Strange that a type so remarkable in its characteristics should persist, in a single representative, so many ages after the period to which it was assigned, to play its part in the wonderful drama of life!

The Trilobites have long since ceased to exist; but afar off, in the Antarctic, science has brought to light a curious Crustacean (*Glyptonotus Antarcticus*), which strongly recalls the extinct form of the Trilobite, as if Nature fondly cherished the reminiscences of her youth. The *Araucaria imbricata* of Chili is, in like manner, a souvenir of the Conifers of the Coal Period, as the Chinese *Salisburia* is of its Sigillarians and Ferns.

Thus, on a review of the history of organic life, we are enabled to draw forth its manifold lessons. We learn that the marshaling of its forms is not in such an order as to justify the fascinating doctrine of genealogical succession, as taught by De Maillet, St. Hilaire, Lamarck, Darwin, Spencer, and others. Still, we learn that order has existed, and that Nature's history may be expressed in formulas. We recognize a bond of thought running through the whole length of creation, and feel the assurance that a Higher Power than physical forces has presided over the evolution of the material world.

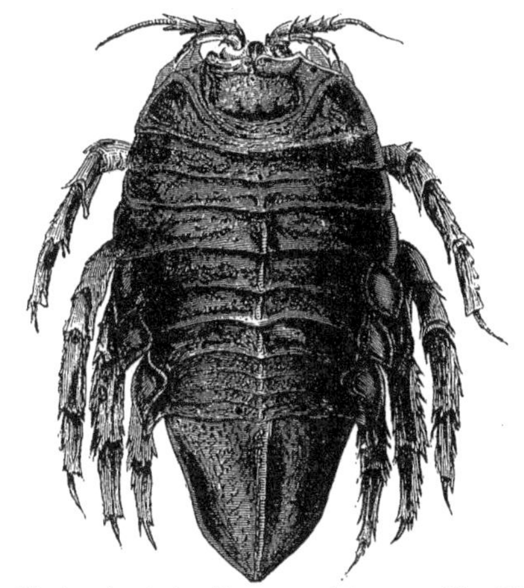

Fig. 94. Glyptonotus Antarcticus, a reminiscence of the Trilobites.

CHAPTER XXIX.

WORLD-THOUGHTS.

LET us trace these world-thoughts further. We shall find the same story thrice-recorded.

Every body knows that the domestic dog outranks the ox, while he is himself outranked by the chimpanzee. Still lower than the ox in rank is that aquatic mammal, the whale, which to every judgment seems just on the separating line between mammals and fishes, while higher than the chimpanzee, both in his organization and his intelligence, is the being, Man. These several forms belong all to the class of mammals, and represent so many orders of that class. Each class of the animal kingdom is composed of different grades of creatures, which mark the different orders of the class. The criteria by which we determine relative rank amongst animals are various. Within the limits of a class, superiority of rank is denoted by aerial respiration, and inferiority by aquatic respiration. Correspondingly, merely aquatic habits, even with aerial respiration, as in whales, show that the affinities are downward. Even the endowment of wings marks a grade below those forms fitted to travel on the surface of the solid earth. Then, again, a multiplication of similar parts denotes inferiority, as is illustrated by the fact that some of the marsupial quadrupeds of Australia possess fifty-four teeth, and some dolphins one hundred and ninety, while the typical number for mammals is forty-four. Thus, also, insects have but six legs, while their inferiors, the spiders, have eight, and myriapods an indefinite number. Inferiority is equally diagnosed by an extension of the abdominal or caudal region, or by a conversion of some of the parts about the head to uses which serve the vegetative rather than the animal functions. Thus the elongated serpent is inferior to the abbreviated turtle; and, in another class, the elongated lobster ranks below the shortened crab.

In each class forms may be selected as ordinal types. In the class of mammals we have *man*, the *monkey*, the *bat*, the *lion*, the *deer*, the *hippopotamus*, and the *whale*, as the expression of so many different ideas in a graduated series. In another class we have *crabs*, *lobsters*, *trilobites*, *lerneans*, and *wheel-animalcules* as a portion of a graduated series of forms expressed under the articulated type. The orders of each class may be regarded as the embodiments of a series of divine conceptions. They constitute a distinct succession of ideas recognizable in a fixed order as the mind glances over the series of organic beings.

Turn now to those wonderful and mysterious evolutions through which every animal goes in passing from the condition of an egg to that of an adult being. We find here expressed the same successions of ideas as in the gradations of adult animals. I have said that aquatic forms stand below terrestrial—the aquatic fishes below the terrestrial reptiles. Now the fish-like tadpole is the embryonic condition of the frog, the toad, and the salamander. In the development-history of these animals, then, the idea of a swimmer and a water-breather is antecedent to that of a land-dweller and air-breather, just as the fish and the whale come before the air-breathing mammal in the ascending grades of being. But what is most astonishing is the fact that *all* vertebrates, *even man himself*, exhibit at one stage of their existence a structural adaptation for the low and fish-like mode of respiration, and by degrees assume the characteristics of higher and higher orders till their destined elevation is reached, when further development is arrested. It is now a favorite doctrine among some embryologists that every higher type, in the progress of its development, passes in succession through phases which represent the fixed conditions of the several orders below it. The author of the "Vestiges of Creation" has consequently undertaken to show at what period of his existence the embryo man corresponds to the fish—at what to the salamander—at what to the tortoise, the bird, the whale, the quadruped, and the ape. Indeed, he goes a step farther, and insinuates that the rank to which any embryo is developed is limited only by the term of incubation or gestation, so that by prolonging this term an offspring of higher grade than the parents may result. There is danger of pushing analogies too far. Similar sequents within certain limits do not warrant us in spurning all limits. Analogies are not to be taken for dependent relations. They may, indeed, express identical plan—identical intelligence—but they are liable to fail at any point. Notwithstanding, it must be admitted that Nature furnishes us in this case with some very suggestive facts. Unlike the author of the "Vestiges," however, I shall employ these facts to show that intelligence presides over creation, instead of proving its absence.

Again, worms are lower in rank than insects. The worm-like grub which cuts off our young corn, and the slugs which eat our cherry and rose leaves, are but the embryos of insects. Here, also, the embryo of a higher type appears under the similitude of the adult form of a lower type. Such illustrations could be adduced at great length. We arrive, then, at the conclusion that Nature, in realizing the succession of phases through which an embryo is made to pass, gives expression to the same succession of ideas as we recognize in the gradations of adult animal forms.

Next, let us recur to the nature of the geological succession of organic types. Every tyro in geological science has learned that we have here, viewed as a whole, a gradually ascending succession of forms. Among other sequences, we find the fish followed in time by the batrachian, which in its embryo state is fish-like, and in its adult state is reptile-like. The batrachian was followed by the strictly air-breathing reptile, in which the ventricle is not separated by a partition—a condition existing in the heart of the embryo mammal. Finally, reptiles were succeeded by quadrupeds and man.

Again, the earliest crustaceans were Trilobites, followed by Phyllopods. These were followed by long-bodied lobsters (Macrourans), and these, in turn, by the crabs (Brachyurans). Now this succession of forms is the same as is expressed in the embryonic history of the highest form of crustacean.

Still, again, fishes with cartilaginous skeletons and (so-called) heterocercal—unequally-lobed—tails predominated in the earlier periods, while our existing waters are tenanted by fishes with bony skeletons and (so-called) homocercal tails. It is a curious fact that this order of succession is represented by the embryonic stages of the common whitefish of Europe, and corresponds also to the discriminations of rank which are recognized in the class of fishes.

Coral animals furnish us with another beautiful illustration of these harmonies. It has been shown by Agassiz, who has enjoyed remarkable facilities for the study of all classes of animals, that the polyps, structurally considered, present a gradation which is expressed, in ascending order, by the following arrangement of groups: *Actiniæ*, *Fungidæ*, *Astræans*, *Porites*, *Madrepores*, *Halcyonoids*. From the *Actiniæ*, whose soft bodies and indefinite multiplication of tentacles mark them lowest in the scale, to the *Halcy-*

onoids, which have but eight tentacles, there exists a regular gradation of complication which in this place can only be announced. It is a fact of extreme interest, however, that the geological succession of coral animals, so far as we can judge, has been coincident with this structural gradation. Of fossil *Actiniæ* we can allege nothing, since, having soft bodies, no relics of them could have been preserved if they existed in the early ages of the world's history. But as to the other groups, we find the *Fungidæ* running through the Silurian, Devonian, and Carboniferous ages, while the *Porites* and *Astræans* bedecked the submarine parterres of Mesozoic Time, and the *Madrepores* united with them to adorn the Tertiary seas, and the *Halcyonoids* belong to the latest times.

A farther extraordinary coincidence is furnished by these coral animals. It has been shown by Agassiz, who examined the Florida Reef under the auspices of the United States Coast Survey, that the true reef-building polyps arrange themselves along the reef in the order of their rank and successive introduction upon the earth. The *Actiniæ* do not appear on the reef for the same reason that they do not figure in the records of geology. They are soft-bodied animals, and never secrete coral. The *Fungidæ*, furthermore, are not compound animals like the reef-builders, and are not confined to any particular depth in the sea. But when we come to the reef-making polyps, we find the true *Astræans* at the bottom, followed by the *Meandrines*, a higher section of the *Astræans*. Next, in ascending along the reef we encounter the architecture of the *Porites*, the *Madrepores*, and the *Halcyonoids* in due succession, and presenting a series of conceptions identical with that found in the structural gradation of polypi, and again in the order of their geological appearance.

What signify now these repetitions of identical successions of ideas? In their structural rank, in their embryonic development, in their geological sequence, and even, in one case, in their relative position in depth, the groups of the animal kingdom give utterance to the same intelligible sentences. This is not the utterance of chance; it is the language of One Intelligence presiding over the evolutions of the organic world through all ages and in all the conditions of its existence.

Other world-harmonies crowd upon our attention. Identical thoughts are written upon the flowers and the stars.

Every one has observed that the leaves of some plants stand in pairs opposite each other, on opposite sides of the stem. In other plants the leaves are scattered over the stem. In this case they are not promiscuously placed, for, on careful observation, we find them disposed in the most regular manner. Commencing with any given leaf, for instance, we shall find the next leaf above this one third of the distance around the stem; the next, another third; and the next another third, so as to stand exactly over the first. The series is, therefore, arranged in a spiral, which may be designated by the fraction $\frac{1}{3}$. Taking another plant, we shall find the next leaf above any given one two fifths of the distance around the stem; the next will be four fifths; the next, six fifths, and so on, each leaf moving two fifths of the circumference farther around the stem. In this case the fifth leaf stands over the first, and this superposition is attained after winding twice around the stem. Here we have an order of arrangement, or a spiral, which may be represented by the fraction $\frac{2}{5}$. In precisely the same way we discover in other plants spirals which may be expressed by the fractions $\frac{3}{8}$, $\frac{5}{13}$, $\frac{8}{21}$, etc. If, in the case of opposite leaves, first mentioned, we conceive that two spirals start from the same level on opposite sides of the stem, it is evident that each successive leaf in each spiral is separated from its predecessor by an interval or angular distance equal to one half the circumference of the stem. We have here, then, a spiral expressed by the fraction $\frac{1}{2}$. The complete series of fractions, therefore, is the following: $\frac{1}{2}$, $\frac{1}{3}$, $\frac{2}{5}$, $\frac{3}{8}$, $\frac{5}{13}$, $\frac{8}{21}$, etc. Now let it be borne in mind that these values are obtained by actual observation, and that there are plants whose leaf-arrangements are known to correspond to each of these fractions severally, as well as others in the series farther continued. But notice the relation which subsists between the successive fractions in the series. Each numerator is equal to the sum of the two preceding numerators, and each denominator to the sum of the two preceding denominators. Knowing this law, we may continue the series to any extent; and it has been so continued, and fractions obtained to which plants have subsequently been found to correspond, though we hardly know how at present to interpret the unrealized possibilities indicated by the higher terms of the series. Is all this the result of chance? Is it not rather mathematics, law, intelligence?

We turn now our attention to the "infinite meadows of heaven," where

> "Blossom the lovely stars, the forget-me-nots of the angels."

Neptune, the remotest planet, revolves about the sun in 60,000 days—speaking in round numbers—Uranus, the next, in 30,000 days, which is *one half* the preceding number; Saturn, the next, in 10,000 days, which is *one third* the period of Uranus; Jupiter revolves in 4000 days, which is *two fifths* of the period of Saturn. And so we go on through the system, and find the law expressing the relations of the revolutions of the planets identical with that which determines the arrangement of the leaves upon the humble stem of a plant.

A little difficulty was at first experienced in applying the law to the group of asteroids; but this difficulty no longer exists, and we now know this wonderful law to be so exact and uniform in its application that, before the discovery of the planet Neptune, the botanist in his garden could have predicted its existence and its place in the heavens with greater precision than the French astronomer in his observatory. Moreover, an examination of this series of fractions renders it impossible that any planet should exist exterior to Neptune, as his periodic revolution corresponds to the beginning of the series; though an indefinite number of planets may exist within the orbit of Mercury, inasmuch as the planets lying in that direction correspond to the indefinite continuation of the series. This correspondence also harmonizes beautifully with the cosmical theory of La Place and Sir William Herschel, which has been explained in the fourth chapter of this work. Astronomers will therefore take notice, and not be found planet-hunting in the deserts of space beyond the orbit of Neptune. In the other direction, the future discovery of an intra-mercurial planet is both possible and probable. Certain it is that presumption sides with Lescarbault in his claim to such a discovery.

Who shall explain what mysterious virtue belongs to the succession of values furnished by the leaf-arrangements of the plant, that exactly the same succession of values should be inscribed upon its humble stem and entered among the ordinations of planetary systems? How many millions of chances against the supposition of a blind coincidence through a series of terms so extended! Premeditated atheism alone could fail to read the sentiment so written at once in the soft bloom of the rose and the supernal light of the stars—"These are the works of one Omnipresent Intelligence."

CHAPTER XXX.

ANTICIPATIONS OF MAN IN NATURE.

TAKING advantage of a midsummer holiday, suppose we visit the country seat of a friend possessed of ample wealth and cultivated tastes. Arriving at the premises, we find the owner called unexpectedly to the city, but the porter, in obedience to the instructions of the proprietor, proffers us a greeting, and bids us in to the enjoyment of the spacious park. We find the grounds laid out and adorned under the guidance of an educated and generous taste. The graveled carriage-road winds under the leafy umbrage of the ancient oaks, or creeps along beneath the dark shadows of a frowning cliff; and ever and anon a sunny opening in the overhanging foliage lets in the golden light upon the quiet-loving Rhododendron and Azalea. Here a modest footpath saunters down a mimic vale, and leads us, worn and weary, to a rustic summer-house all overarched with honey-breathing Loniceras intertwined with the scandent Cobea and woodland-loving Bignonia. Here are seats provided for the languid visitor; and from the roots of the thirsty beech, whose overreaching branches rib the leafy arch, bursts forth a laughing fountain, while a goblet standing by seems to say, "Here the visitor will be thirsty and warm, and will eagerly refresh himself at the cooling spring." The proprietor of the grounds, though not here in his visible presence, has left here the evidences of his thoughtfulness and expectation of a wearied visitor. Then for the first time we spy what is equally welcome with the cool fountain—a basket of ripe and luscious fruit, revealing itself, like Heaven's blessings, just at the moment when nothing could be more desired. How well the owner of the premises knows how to minister to the wants and pleasures of his guests! Refreshed, we wander on through a darkly-shaded copse, when a sudden elbow in the foot-path brings us to the rock-built doorway of a rustic grotto. The cool lintels are hung with brown and emerald fringes of dew-dripping mosses, and the leaf-strewn portal of the dusky hall reminds us of the cave of the Cumæan Sibyl. The desire to enter this enchanted grotto has been foreseen by the care which provided a flight of half a dozen steps, down which we descend to the damp, chill floor of the Sibylline abode. The long, dim hall before us fades into indistinctness in the distance, like the line of memories receding toward childhood's years; and just as our timid steps are about to be reversed, we espy some matches and a taper resting on a shelf of rock, and, with the light so opportunely provided, explore the length of the charming little cavern.

Emerging from our subterranean exploitation, we visit, in succession, all the remaining nooks and surprises of the well-plotted grounds, and find that every where the thoughtfulness of the proprietor has preceded us, and ushered our coming with the most intelligent preparation. Not the least admirable of the arrangements of his shrewd forecasting is his occasional combination of geometrical figures cut in the turf of a growing grass-plot, or traced in the airy edgings of the most exquisite flower-beds, themselves the achievements of geometrical skill, and adapted specially to please the mind and fancy trained in mathematical forms. The work itself bespeaks a skillful mind, and equally proclaims an expectation of educated guests. This lavishment of learned conceptions is not the mere gamboling of genius for its own amusement. What we see, and enjoy, and comprehend declares in plainest language not only that the contriver of these grounds possessed superior intelligence, but that he expected intelligent guests to visit, admire, and enjoy them.

This admirably plotted park is the domain of Nature. These dark, umbrageous shades and quiet dells are hers. These winding highways and meandering footpaths are her navigable streams, and lakes, and ocean tides. The rhododendron and azalea were first planted by the hand of Nature, and her fingers taught the honeysuckle to climb the rustic trellis of oaken boughs. Her providence drew forth the crystal fountain beneath the beechen shade, and her foresight laid by the store of coal with which we warm and light our dwelliugs.

To be more specific, let the reader imagine that the history of the world had been a scene of never-ending quiet. Suppose a fear of inflicting animal suffering had laid its restraining hand on the volcano and the earthquake; suppose the rocks had not been plowed up, and the deep subsoil of the earth's crust laid over in mountain ridges. I do not ask whether, in the midst of scenes of such monotony, the occasion could ever have arrived for the deposition of the coal. We will assume that it would. I do not ask whether, without eruptions and terrestrial distresses, the precious and the useful metals would ever have been reduced from their ores; we may assume that they would. But where would lie our coal? Buried ten thousand feet from view, man would never have learned of its existence, much less would he have known how to raise it to the surface. See the provision of Nature in breaking up the coal-bearing strata and tilting them on edge, as much as to say, "Lo! here is your desire; search not in vain; dig, and be satisfied with warmth; drive forth the hidden energy of the abundant water, and bid the servants furnished to your hands execute all the behests of your convenience." Had chance formed the beds of coal under such a concurrence of auspicious and beneficent conditions, chance would not have brought it to our doors; chance would not have rescued it from burial beneath the sediments of a thousand following ages; chance would not have laid by in the same beds the ores of iron which the coal is fitted to reduce; chance would not have stored in the same relation the beds of limestone, to be used as a flux in the reduction of the iron ores by means of the mineral coal. See what provident Nature has done with other metals! Was it accident that enriched the upper peninsula of Michigan with her wealth of native copper? Or has there been in existence upon our earth any other being than man to whom these riches possessed the least utility or interest? The ores of copper lie buried a mile beneath the sandstones of the "Pictured Rocks." The sediments of unknown cycles of years were gathered upon the beds of valuable ores. At length, while the world was preparing for man, a fiery outburst threw the deep-buried treasures to the surface. It did more. It reduced their refractory ores for the hand of man, and enabled him to gather directly the native metal. Still more. The same fiery outburst bent the flinty rocks into the form of a huge trough, and Heaven sent down water to fill it and float the steam-sped vessel to the copper-bearing shore. And lastly, lest the manhood of our race should be spent before the discovery of the treasure, all-provident Nature broke up samples of cupriferous rock, and strewed them along the shore, and along the river-courses, so that, when man should find them, he might trace the trail, as by a clew, to the original store-house of the native metal. And all these preparations, and provisions, and utilities have no relations to any other terrestrial denizen than man.

Nature has mined for us in gold. Deep in the rocky recesses of the earth lay the precious metal. It must be brought to the light of day. But Nature does not do this till the work of sowing sediments—the seeds of rocky growth—has been completed over all the areas destined to be inhabited by man. Had the deep-treasured gold been brought up in the Mesozoic Ages, the inundations and vicissitudes of later times would have scattered it over the breadth of the land and the sea before our race had made its advent. No such false step was taken. It is only after the Tertiary beds have been all deposited that Nature throws up innumerable veins of quartz, which bring along with them the glittering gold. This is well; but Nature possessed a quartz-crushing machine in the shape of a glacier a mile in thickness, and some hundreds, if not thousands of miles in horizontal extent, and this she drew over the projecting veins of auriferous quartz and ground them to powder. These, at least, are the general views put forward by Sir Roderick Murchison in regard to the principal gold regions of the world. The California geologists, however, aver that the great ice-plow never scored the ribs of the Sierra Nevadas. Nature may have pulverized the gold-quartz of our Western states and territories by some other agency. Nevertheless, it has been crushed and comminuted on a stupendous scale. When this work was done, by whatever means, she brought her gold-washing machine into requisition, and "jigged" the golden sands till the yellow particles were well assorted, and then strewed them along the narrow ravines to await the attentions of the coming man.

But we need not go to the golden sands of the Sacramento to read the anticipations of man in the arrangements of Nature. What is every well and spring but a subterranean stream that has been beguiled to light by the outcropping of the impervious floor over which it had flowed? We need not attempt to imagine what would have resulted were the rocks left to rest in horizontal and continuous layers, but it is worth while to recognize the beneficence of that vast accumulation of loose materials which we call drift. It is, as it were, an enormous sponge, which drinks in the showers of heaven, and stores them away beyond the reach of defilement and putrefaction in the deep, cool reservoirs of the filtering sand-beds, so that it is almost impossible to penetrate the drift to a depth sufficient to secure an agreeable coolness without obtaining a plentiful supply of well-strained water. So common and so vital a comfort has been secured by the geologically-extraordinary deposition of such masses of loose materials over the surfaces of the naked rocks, and not less by their distribution in beds of sand and clay presenting every possible irregularity of thickness, extent, and disposition. (See Fig. 84.)

These and multitudes of other arrangements, collocations, structures, and products of a useful and beneficent character, are so many indications that during the long process of the world's fitting up—while yet the human era was contemplated as we contemplate the millennium—man, the nature of man, and the wants of man, constituted at least one of the objective points of cycles of geological preparation.

Finally, it is eminently worthy of remark that Nature has not only anticipated the coming of man, but has contemplated the exercise of human intelligence. How few of the benefits which Nature affords have been reached without study and thought! None will affirm that matter was endowed with all its capabilities of benefit to the human race without any design that those benefits should be secured and enjoyed. This is tantamount to saying that the provisions of Nature prophesy a reasoning mind. We may venture to go much farther than this, and assert that the material of thought which Nature furnishes is correlated to the thinking principle in man. When the Creator adopted an intelligent method in the ordinations of the material world, it was equivalent to a declaration of purpose to introduce an intelligent being. And when the Creator had stocked the world with the materials of thought, and had planted in it a being capable of understanding Nature, it was the obvious purpose of the Deity that Nature should be investigated, and that, by such investigations, man should become not only wiser, but more reverent, religious, and happy.

CHAPTER XXXI.

THE TOOTH OF TIME.

A FEW words about the disintegration of the rocks. As the vital force employs itself in the demolition of the organic structures and the simultaneous repair of all the wastages, so the gigantic energies of geology have busied themselves in one age or place in demolishing the rocky fabrics consolidated with incredible labor in another age or place. The grain of sand upon the rivulet's border may have been incorporated successively into a dozen different formations, each in turn disintegrated to be inwrought in the rocky sheets of the next succeeding age.

Has the reader ever inquired whence came the materials for twenty-five miles of sedimentary strata? It is a question which geology is compelled to answer. The first and lowest great system of strata—the Laurentian—is in Canada thirty or forty thousand feet thick. This system is supposed to embrace nearly the entire globe, passing beneath the Paleozoic, Mesozoic, and Cenozoic strata, and extending, probably with greatly diminished thickness, under the beds of the existing oceans. It must have been accumulated while yet the primeval sea was wellnigh universal. This is the prevalent opinion. It is perfectly plain, however, that these vast beds of sediment must have had an origin in pre-existing rocks lying within reach of the denuding agencies of the time. How enormous a bulk of solid rocks was ground to powder to furnish material for these Laurentian strata may be imagined when the reader is reminded that the mean elevation of North America is but about twelve hundred feet above the level of the sea; and if the entire continent were ground to powder down to the sea-level, and distributed over an area of the ocean's bottom equal only in extent to North America, it would afford a bed of strata not one twentieth the thickness of the Laurentian system over the same region. Whence, then, the materials for so vast an accumulation of sediments? Where were the lands which must have disappeared during the Laurentian Age? Although we may not be able to indicate their location, the facts suggested serve to remind us of the gigantic scale of operations of the denuding agencies of primeval time.

Every succeeding geological age must also have had its source of supply to the contemporaneous sediments. The ever-growing continents were ever wearing down. As the increasing pressure of the accumulating oceans crowded higher the summits of the continental axes, the ceaseless demands of the insatiate sea for more sediments wore thinner and thinner their denuded scalps. It is no wonder that included fires burst forth at the summits of the highest mountains. These are the exposed points, where the earth's crust has been reduced to the greatest degree of tenuity, while the ocean's floor is the most solid portion of the globe.

The outburst along the southern shore of Lake Superior at the close of the Potsdam period developed topographical features of infinitely greater ruggedness than those which now characterize that region. Kewenaw Point, the Porcupine Mountains, and the Huron Mountains, as well as the numberless unnamed knobs still standing throughout the region, have been gnawed and battered down for hundreds of feet, and their once angular outlines have been scoured to a subdued rotundity. The Appalachians, that once lifted their multiplied folds to the heights of the An-des, have been planed down to the level of third-rate mountains, and the dust and rubbish scraped from their worn heads has been deposited in the troughs between the ridges, or strewn along the coast to form the foundations of the Atlantic States.

The city of Cincinnati stands upon the exposed core of the broadest and most westerly of these Appalachian folds. The rocky wrappings of this axis have been planed down from the summit till the Carboniferous, Devonian, Upper Silurian, and Lower Silurian strata have been successively reached, and these superincumbent layers tilt in all directions from the summit of the arch. The graves of encrinites and brachiopods, that had lain undisturbed for untold geological cycles, buried a thousand feet beneath ocean-slime and careering waves, and, at a later period, beneath the roots of Carboniferous tree-ferns and the mire of steaming jungles—graves of populous nations of the olden time—were uncovered and plowed from their locations, and crumbling skeletons were strewn over the area of three states. So the spirit of modern improvement sometimes lays its remorseless hand on the cemeteries of man, and sunlight falls again upon relics that had once been laid away with sacred care. What veneration fills the mind of the geologist as he walks over this waste of the Silurian burial-vaults. The Ohio has plowed its way through the buried city, and a city of the living has been reared upon the bones of the dead. The native ramparts which wall in the Queen City upon every side are but the broken tiers of vaults in this ancient cemetery. The area of this desecration extends to Madison and Richmond in Indiana, and to Danville and Richmond in Kentucky. Throughout this entire extent of country the once superincumbent formations have been swept away, and the material wrought into the structure of formations of later date.

I ascend to the cupola of the magnificent state-house at Nashville, and take a survey of the surrounding country. On every side spread out the broadly undulating fields of grass and corn into the illimitable distance. A finer agricultural scene was never witnessed. A more beautiful landscape, diversified with broad clearings, waving crops, tufts of magnolia and poplar, shining mansions, withdrawing vales, and purple atmosphere, it has never been my privilege to gaze upon. What is the substratum of all this beauty of form and landscape? Descending to the ground, I find myself standing again upon the opened sepulchres of Lower Silurian populations. I go down to the bank of the Cumberland, and view the sharp-cut walls which frown above the muddy current a hundred feet below. Here is a deep perpendicular gorge chiseled by the river through the marble strata of the Trenton and Cincinnati groups. I set out upon an exploration of the charming country mapped before me from the dome of the Capitol. Traveling eastward for sixty miles, I pass continuously over an undulating exposure of the same strata. Here I find an outer wall four hundred feet high, which bounds this magnificent basin of Middle Tennessee on every side (Fig. 95). I climb to the top of this wall, and ascertain that it is at this point, the western termination of a series of overlying strata of Silurian and Devonian age, which to the west have been swept away, but toward the east form an elevated plateau, through which the streams have scored deep gorges four hundred feet down to the level of the central basin.

This "highland rim," as my scientific friend, Professor Safford, styles it, is forty miles wide. We come then to the foot of the Cumberland Mountains—or, more properly, Cumberland Table-land—and ascend a thousand feet over the outcropping edges of Lower Carboniferous strata, and

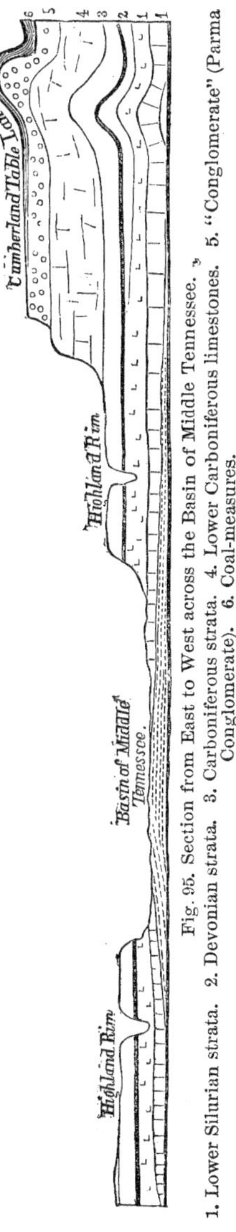

Fig. 95. Section from East to West across the Basin of Middle Tennessee.
1. Lower Silurian strata. 2. Devonian strata. 3. Carboniferous strata. 4. Lower Carboniferous limestones. 5. "Conglomerate" (Parma Conglomerate). 6. Coal-measures.

find the brink of this higher wall composed of beetling escarpments of the Coal conglomerate. I mount one of the overhanging cliffs at Bon Air—an old but now ruined watering-place and summer resort—and look down hundreds of feet upon the green tree-tops, from whose sunny summits ascends the chorus of a myriad warblers. Far away to the west stretches the landscape over which I have traveled, and its farther verge blends with the azure which overarches all. Far toward the north and south spreads out the basin of Tennessee; and over all hangs the purple haze with which Nature half conceals, and thereby heightens her charms. Southward rise sheer from the plain some isolated knobs and ridges, which mark the commencement of the region whose general structure embraces Lookout Mountain, Missionary Ridge, and similar precipitous elevations along the southern border of the state.

From the dizzy brink on which I stand stretches eastward a cool and salubrious table-land, known as the Cumberland Table-land, marked by a soil, and forest-growth, and climate as distinct from those of the basin below as Wisconsin from the prairie region of Illinois. The underlying strata are nearly horizontal. Why now do they terminate so abruptly at the mountain wall which we scale to reach Bon Air? The cut margins of these mountain ribs lie exposed and protruding from base to summit of the laborious ascent. Did Nature form them originally thus? We are forced to conclude that these mountain sheets, like those under the "highland rim," once extended westward over the basin of Tennessee, and have been scooped out by some tremendous agency appointed by Nature to furnish materials for the states—then future—of Georgia, Alabama, and Mississippi. The Cumberland Table-land and the abrupt knobs about Chattanooga are not upheavals, but lines of relief. It is the valleys that have been made, and not the mountains. The mountains are landmarks—"benchmarks" the engineer might say—showing the former level of the entire region.

If we travel westward or northward from Nashville we find the basin walled in by the "highland rim," though it is only on the east that the pile of strata rises so high as to bring us within the limits of the Coal-measures. Here, then, is one of the most stupendous examples of geological denudation. What the precise nature of the agency by which this work was done we can only conjecture. Equally uncertain is the precise date of the work. We can only say that it was performed between the close of Paleozoic Time and the present, but as to the reality and almost incalculable vastness of the work we have no room to doubt. Neither can we fail to see that such enormous excavations must have been in progress in all ages, to furnish the requisite amount of materials for formations of continental extent, and attaining a thickness of hundreds or thousands of feet.

I have already alluded to the monuments of destructive

action around the shores of the great lakes. Even mimic oceans like these, in the era of their strength, have performed labors which excite our astonishment. And that Titanic power which geology dimly pictures to us as moving in glacier-masses from parallel to parallel, riding over primeval forests, obliterating ancient river-beds, plowing out lake-basins to the depth of nine hundred feet, and crushing to powder countless cubic miles of obdurate granite and quartz—that power of which we can little more than dream, though the records of its marvelous march are scattered about on every side—a power which may have been summoned into exercise at more than one period in the world's history—that power whose movement was resistless as fate, and destructive as the crash of worlds, can serve at least to impress our minds with the energy of geological agencies, and the resources at Nature's hand for the scooping of lake-basins, the carving of mountain cliffs, or the scraping out of the bowels of the State of Tennessee.

Even the humble river-stream—humble by comparison, but terrific as Niagara in unwasting and untiring power—has accomplished work at which the highest human engineering stands appalled. The Kentucky and the Cumberland, in traversing the states which they drain, have worn their channels to the depth of hundreds of feet below the general level of the country. Some of the wildest and most attractive scenery of the continent lies along the Kentucky, from the mouth of Hickman's Creek in Garrard County, to Dix River and Coger's Ferry in Mercer County. Even the smallest streams have aped the pretentious labors of the larger, and have succeeded in opening their narrower gorges through two, three, and four hundred feet of the blue limestone of the blue-grass lands of Kentucky and Tennessee.

But these all are pigmy works compared with those of

the streams which traverse the "Great American Desert." For the most vivid descriptions of the geology of this forsaken region we are indebted to Dr. Newberry, the geologist of Ives's Colorado Expedition under the general government. The surface formations are mostly of later Carboniferous and Mesozoic age, interrupted at intervals by mountain-like outbursts of volcanic origin. The region is a vast plateau stretching for hundreds of miles in either direction (Fig. 96). The floor of the plateau is a mass of horizontal strata. Far in the hazy horizon may be seen the bold wall, which rises to a more elevated table-land composed of overlying strata. These higher strata were once continuous over the surface of the lower plateau, but have been swept off by denudation. Still farther in the horizon looms up another gigantic terrace, rising to the upper plateau of the desert. The traveler journeying across this apparently monotonous and desert plain finds himself suddenly standing on the brink of a precipice. It is the wall of a deep gorge. Down into this gloomy chasm he endeavors to cast a look. It is like a vertical rent through the strata to the appalling depth of more than a mile. Far down at the bottom winds the sky-lighted stream which has executed this tremendous piece of engineering, quiet now as a lamb, but in spring-time roaring and destructive as a lion. This is the Colorado. Its immediate banks are fringed at intervals by a narrow border of grass, and these meagre grass-plots down in the rocky cleft are the occasional abode of the desert Indian. The great Black Cañon of the Colorado is a gorge with perpendicular walls of rock three hundred miles long, and from three thousand to six thousand feet high! The lateral streams have cut similar gorges, and these almost impassable chasms constitute formidable difficulties in traversing the country (Fig. 97). The Colorado has cut through the entire series

Fig. 96. Upper Cataract Creek, near Big Cañon, Colorado.

of formations, and sunken a thousand feet into the solid granite. The section of the rocks in the gorge shows above the granite two or three thousand feet of paleozoic sandstones, shales, and limestones, one thousand feet of sub-carboniferous limestones, and twelve hundred feet of carboniferous sandstone and limestone. Higher up the stream the section extends up through the Triassic and Cretaceous systems.

What æons have rolled by while this unparalleled river-work has been in progress! And yet this work must have been limited to the later ages, since the gorge cuts through Cretaceous strata. There was a time, during the Cenozoic ages, before yet the ridges of the Rocky Mountains had been elevated to their present altitudes, when this vast desert had just become dry land—upheaved from the recent bottom of the Cretaceous sea. Now the Colorado began to gather its forces and to irrigate the surface of the new-formed land. Now began the great cañon; but for many ages the surface features of the region were normal; and not improbably it was clothed with a soil, and watered by streams which sustained a luxuriant growth of vegetation. But man was slumbering in the voiceless future, and lazy reptiles held possession of the fair domain. Vast, then, as is the work, and vast as must have been its duration, its commencement can date back but to the end, or, at farthest, to the beginning of Cenozoic Time.

Who can tell but similar gorges have been cut in the strata of more eastern states. Here was land—permanent land—covered with vegetation, while yet the great desert was but ocean-slime. Here, too, were rivers—rivers like the Ohio and the Mississippi—with their numerous tributaries. What prevented these streams from scoring the strata to the depth of ten thousand feet? We know that during this interval the Niagara cut an ancient gorge. We

Fig. 97. "Big Cañon" of the Colorado.

know that an ancient river-bed stretches from Lake Michigan down through the valley of the Illinois. The subterranean explorations of the well-borer's auger have disclosed multitudes of ancient gorges which are now filled up with drift. If such tremendous gorges were ever cut, they were filled up and obliterated by the great glacier. And may not this reparation of the surface have been one of the beneficent operations of the glacier? We are told no glacial action is detected west of the Rocky Mountains. Had the great glacier been moved over the deep-cut gorges of the great desert, they must have been filled and blotted out, and the new-formed streams, on the advent of man, would have been just in the act of surveying new channels for themselves. The bare rock would have been clothed with soil, and the "desert" might have been the garden of the continent.

CHAPTER XXXII.

PRIMEVAL MAN.

THE history of our race, traced back a few thousand years, loses itself in traditions and myths. We come down out of a cloud of obscurity, in which we can just discern the rude forms of men clad in skins, frequenting the caves of wild beasts, fashioning rude pottery, and practicing in the chase with the primeval bow and arrow. Out of the haze which hangs over the verge of antiquity come sounds of conflict in arms, pæans of peace, hymns to religion, and the hum of barbaric industry.

Our written history does not extend back to the origin of man. The Mosaic records, which are undoubtedly the oldest of our authentic documents, represent the western portion of Asia as swarming with a population tolerably advanced in the arts at a period two or three thousand years antecedent to our era. There was, consequently, a long interval of human history still anterior to this date. What destinies befell our race—how did they live, whither did they wander, during that prolonged infancy of which—Revelation aside—we have no other information than such as we have gleaned of the Mastodon, the Megatherium, or the Zeuglodon?

The quickened intellectual activity of the modern age has started new and interesting inquiries in this direction. There are no questions which more profoundly interest us than the history of primeval man. The investigation has been pushed far beyond the limits of the most ancient written documents. It has passed over the remoter domain of

Fig. 98. Pre-historic Man.

archæology and stepped upon the ground consecrated to the researches of geology.

The chief sources of our information respecting the earliest periods of human history are, 1st. The remains of man himself, which have been found in caves or buried in deposits of gravel or peat. 2d. Human works, of which we have the so-called Druidical remains of Great Britain and other countries, known as dolmens or cromlechs—rude megalithic monuments of unhewn stone, which we now know to be ancient tombs. Other human works more abundant and more universally distributed are implements of war, of the chase, of industry, or of ornament. These are found in gravel-beds along the valleys of rivers or at their mouths; in peat beds; in caves, and among the refuse piles contiguous to the camping or dwelling-places of tribes which subsisted partly upon molluscs. These refuse heaps are composed mostly of shells of recent species, bones of domestic or wild animals suitable for food or service, fragments of pottery, arrow-heads, fish-hooks, stone implements, ornaments, and the like. A vast supply of the relics of primeval man has been obtained from the pile-habitations, or ancient dwellings constructed upon platforms supported by piles driven in the water. The dredging of the bottoms of these lakes has brought to light immense quantities of the remains of pre-historic art and industry. 3d. The manner in which the relics of man are associated with those of other animals enables us to extend to our race many of the generalizations deduced in reference to the earlier history of the existing fauna. Lastly, the nature and magnitude of the geological changes which have transpired during the existence of man throw some light upon the antiquity of the race.

As in the history of organic life in general, so in the geological history of man, we find him mounting from lower to higher manifestations in the progress of the ages. There seems, however, to be a fundamental difference in the two kinds of progress. With the lower animals it is a structural advance; with man, an education. With the former the steps of the advance are marked by successive species; with man by successively higher attainments of the intelligence. With the other vertebrates the highest is structurally different; with the succession of human races, the highest and the lowest are structurally identical.

Archæologists distinguish three ages in the history of man—the Age of Stone, the Age of Bronze, and the Age of Iron. In the Age of Stone, the uses of the metals had not been discovered, and human implements were constructed of flint, serpentine, diorite, argillite, and other suitable rocks. In the Age of Bronze, implements of bronze began to be introduced, and we descend to the verge of historic times. The Age of Iron is characterized by the use of that metal, and the arts and industries of the most advanced civilization.

Most anthropologists are inclined to subdivide the Age of Stone into two or three epochs. Vogt, Lartet, and Christy divide it into two: first, the Cave-Bear Epoch, or epoch of hewn stone implements; secondly, the Reindeer Epoch, or epoch of polished stone implements, carved and artfully decorated bones, and other evidences of "a very intelligent, art-endowed race of men."

It is not by any means certain, however, that these two epochs were successive. The more skilled workmen of the Reindeer Epoch may have lived contemporaneously with the Cave-Bear men, as natives of all degrees of civilization have co-existed upon the earth in all ages. Neither is it supposed that the three ages represent three stages of human civilization, each of which, in turn, has been worldwide. We find simply that in the history of every race there is a Stone Age; and if the race advances, this is followed by an Age of Bronze, and this by an Age of Iron. Some Eastern nations passed out of their Stone Age three thousand years or more before the Christian era. Some of the peoples of Central and Northern Europe were in their Stone Age when Cæsar subjugated Gaul. The Sandwich Islanders were in their Stone Age when first visited by Capt. Cook, while the Esquimaux and the North American Indians generally are still in their Stone Age. The Age of Stone is simply the stage of infancy. Different peoples have emerged at different epochs from the state of national infancy.

When man first made his advent in Europe, that continent was still the abode of quadrupeds now long extinct. The contemporaries of man in the Hewn-stone epoch were the Cave-Bear (*Ursus spelæus*), followed by the Cave-Hyena (*Hyena spelæa*) and the Cave-Lion. These gradually gave place to gigantic herbivores—the Hairy Mammoth (*Elephas primigenius*), the Hairy Rhinoceros (*Rhinoceros tichorinus*), and the Reindeer. The mammoth roamed in herds over the whole of Europe, Northern Asia, and even North America. The hairy, or two-horned rhinoceros, in company with another two-horned species, thundered through the forests, or wallowed in the jungles and swamps. The rivers and lakes of Southern Europe were tenanted by hippopotami and beavers—the former as huge and unwieldy, and with tusks as large as any which terrify the African Bushman. Three kinds of wild oxen, two of which were of colossal strength, and one of these "maned and villous like the Bonassus," grazed with the marmot, and wild goat, and chamois upon the plains which skirt the Mediterranean. The musk-ox and the reindeer browsed in the meadows of Perigord, in the south of France, while a gigantic elk (*Megaceros hibernicus*) ranged from Ireland to the borders of Italy (Fig. 99).

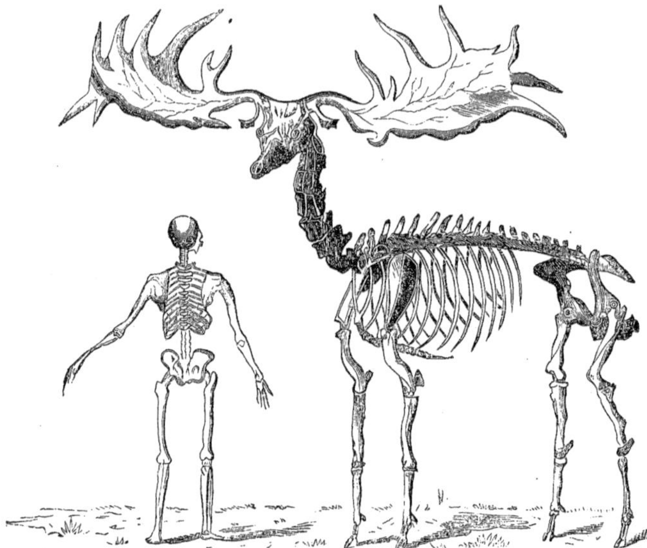

Fig. 99. Skeleton of Extinct Giant Elk (*Megaceros hibernicus*) of Ireland, compared with Man. (Reduced from an Irish lithograph.)

That these animals lived as contemporaries of man is proven by two classes of evidence. In the first place, the bones of man and the relics of his industry are found preserved in the same situations as the bones of these extinct quadrupeds. In 1828, Tournol and Christol disclosed the coexistence of such remains in the caves of the south of France; and, somewhat later, Schmerling described from caves in the environs of Liège, bones and even crania of men, together with arrow-heads and other articles enveloped in the same stalagmites with the remains of the mammoth, rhinoceros, cave-bear, cave-hyena, and other animals. A similar association of remains has been observed by Austen in the celebrated cave of Kent's Hole, near Torquay, in England. More recently still, more important discoveries have been developed by M. Lartet from the caves in the south of France. In 1841, M. Boucher de Perthes published to the world an account of human remains found buried in the valley of the Somme, near Abbéville, in company with the bones of extinct species of quadrupeds. In 1842 M. Melleville reaffirmed these discoveries, and in 1844 M. Aymard presented new facts disclosed by explorations upon the slope of the mountain of La Denise, near Puy. In 1853 Dr. Rigollot announced the discovery at St. Acheul, near Amiens, of hatchets and articles in cut stone, found imbedded in the same gravel deposit with the fossil remains of the hairy elephant, rhinoceros, and extinct ox. Similar discoveries have been reported from Spain, Italy, Greece, Syria, and England. In the United States we detect also some evidences of the coexistence of man and extinct species of quadrupeds. Dr. Koch, the reconstructor of the Tertiary Zeuglodon, insisted long ago that he had found in Missouri such an association of mastodon and Indian remains as to prove that the two had lived contemporaneously. I have myself observed the bones of the mastodon and elephant imbedded in peat at depths so shallow that I could readily believe the animals to have occupied the country during its possession by the Indian; and gave publication to this conviction in 1862. More recently, Professor Holmes, of Charleston, has informed the Academy of Natural Sciences of Philadelphia that he finds upon the banks of the Ashley River a remarkable conglomeration of fossil remains in deposits of post-tertiary age. Remains of the hog, the horse, and other animals of recent date, together with human bones, stone arrow-heads, hatchets, and fragments of pottery, are there lying mingled with the bones of the mastodon and extinct gigantic lizards.

Contemporary with these American animals, but not yet found associated in their remains with the relics of the human species, lived, in North America, horses much larger than the existing species, grazing in company with wild oxen, and herds of bisons (*Bison latifrons*), and shrub-loving tapirs (*Tapirus Americanus*). The streams were dammed by the labors of gigantic beavers (*Castoroïdes Ohiensis*), while the forests afforded a range for species of hog (*Dicotyles*), and a grateful dwelling-place for numerous edentate quadrupeds related to the sloth, but of gigantic proportions.

In the next place, evidences of the contemporaneity of man with species of quadrupeds now extinct are found in carved and deftly-fashioned implements and other articles made of the horns, bones, and teeth of these animals, and especially by representations of the outlines of many of them executed upon ivory, bone, horn, and slate. The most remarkable discoveries of this kind have been made by M. Lartet, in 1864, in the caves of Perigord, in the south of France. In the midst of the soil and *débris* with which the bottom of these caves is covered have been exhumed various etchings of animals, executed on pieces of the horns of the deer and the ivory of the elephant. One of these sketches represents a deer, one the head of a wild goat, another an elk allied to the moose, another the head of a reindeer, another the head of a wild boar, and still another nearly the entire outline of the hairy mammoth (Fig. 100), which conforms marvelously with the restoration of this proboscidean published by the Russian naturalist Brandt. There can be no question but that the artists were personally acquainted with the animals which they outlined (Fig. 101).

As we descend to the epoch of the Reindeer folk, the principal change in the fauna of Europe consists in a diminution of the number of carnivores and an increase of the ruminants. The mastodon, elephant, reindeer, elk, and oth-

Fig. 100. Profile of a Hairy Mammoth engraved on a piece of elephant's ivory, found in the Madelaine Cave, France.

er large ruminants still survived. Messrs. Christy and Lartet found a vertebra of a young reindeer transfixed by a flint arrow-head. Ornaments made of teeth have been discovered, and from several caves have been obtained bone whistles, formed by boring the carpal and tarsal bones of ruminating quadrupeds. A sculptured dagger, made of a single piece of reindeer's horn, attests the contemporaneous existence of that animal in the south of France. A cylindrical tool found in the same vicinity bears upon one side the heads of two aurochs, while upon the other are the profiles of two horses, with a human face between them.

Fig. 101. The Hairy Mammoth (*Elephas primigenius*) restored.

The exploration of the dolmens, or monuments of enormous unhewn stones, so abundant in France, England, and Scandinavia, but unmentioned in the most ancient history of these countries, shows them to have been constructed by men of the Reindeer Epoch. Some of them, from the presence of polished stone and even of bronze implements, belong evidently to the closing stage of this epoch. These megaliths have been found not only in the regions anciently inhabited by the Celts, but also in Syria, Africa, and even in Hindostan.

A similar association of human relics with the bones of quadrupeds occurs in the turf-pits of Denmark, and the *Kjœk kenmœddings* of Denmark and Sweden. The only extinct animals recognized in the latter are the lynx and urus, though bones of the hog and dog are also common.

To the latter part of the Reindeer Epoch belong also the pile-structures discovered in the lakes of Switzerland. The only extinct species are the elk, the aurochs, and the urus. Remains of still-existing species, as the brown bear, the badger, the pole-cat, the otter, the wolf, the dog, the fox, wild-cat, beaver, wild boar, goat, and sheep, exist in great abundance in the *débris* dredged from the bottoms of these lakes.

Of the animals thus shown to have lived contemporaneously with primeval man upon the continent of Europe, the cave-bear, cave-hyena, tiger, mammoth, mastodon, and others of less importance became extinct before the date of written history; but these extinct quadrupeds had lived contemporaneously with others which have come down to historic times. The reindeer, referred to by Cæsar in his Commentaries, is thought to have survived in Northern Scotland as late as the twelfth century; the Irish elk existed up to the fourteenth century; the reindeer continued in Denmark till the sixteenth century; the urus lingered in Switzerland up to the sixteenth century; the bison still survives in Lithuania, and the wild boar is abundant in Central Europe.

It is commonly supposed that the Reindeer folk were the successors of the Cave-Bear folk; but Dr. Packard has very plausibly suggested that they may have lived contemporaneously, side by side. "The Reindeer folk may have inhabited the upper valleys and hills near the Alps and Pyrenees, which send spurs into Southern and Central France. They were perhaps mountaineers, and the animals associated with them and most characteristic of the period were Alpine and northern species. * * * Their neighbors, the Flint folk, or Lowlanders—a taller and stronger race—meantime inhabited the plains of Northern France and Belgium, England, and Germany, and the fauna was made up of the mammoth, mastodon, rhinoceros, horse, cave-bear (which was more abundant than with the Reindeer people), bison, aurochs, and deer, which inhabited the more genial and fertile plains."

The geological status of the continents on man's first appearance was unique. They had just emerged from the reign of ice. The glaciers had begun to retreat, but, except in Southern Europe and Middle Asia, the climate was still rigorous. The hairy elephant and rhinoceros, clad in winter furs, as well as the fur-clad bear and hyena, found a fitting abode upon the shores of the Atlantic and Mediterranean. The marmot, the wild goat, and the chamois, now confining themselves to the cold peaks of the Alps and the Apennines, lived then upon the lowlands of France and Spain. The musk-ox, in our day restricted to the regions beyond the sixtieth parallel of latitude, grazed in the cold marshes of Dordogne. On the American continent, the subsidence which terminated the reign of frost was not arrested till a large portion of the United States had been again submerged; and on the Oriental continent the indications of northern depression are equally unmistakable and equally extensive.

The moment that the last revolutionary visitation had come to an end—while yet the lands had become scarcely stable in their places—man seems to have suddenly made his appearance among the beasts of the earth, and to have moved among them and controlled them with a conscious and uncontested superiority. Let us see what can be learned of the habits and endowments of this primeval man.

Was primeval man created in Europe, where we have the earliest traces of his existence, or was he here an emigrant from the East? In answer to this question we can produce no decisive facts. There are, however, considerations of weight. In all the later epochs, even of the Age of Stone, there was evidently a continuous migration from the direction of the Asiatic hive. The movement of population has always been westward in regions to the west of the Orient, and it has always been eastward in regions to the east of the Orient. The westward wave overflowed Europe, and in later days crossed the Atlantic. The eastward wave populated Tartary and China, and, as may be presumed, dashed across the Straits of Behring, and flooded the American continent at a remote period. To say the least, till the American shores were reached by the westward wave from Europe, the tide of population in America had always set from north to south. The primeval inhabitants of North America were Asiatics in their features, their language, and their arts, and tradition speaks of them as moving from the direction of Asia. These movements of human populations, like radiating streams, from the western part of Asia, certainly afford a presumption that the only people of whose movement we have neither history, tradition, nor buried monument, proceeded also from the direction of the Orient.

From the same quarter of the world proceeded most of our domestic animals and plants, and in the same quarter of the world the perpetually uttered prophecies of the geologic ages proclaimed that the line of animal life should have its culmination. We have, then, strong presumptive evidence that the men of the Stone Age were brethren of the men who came afterward from the East and taught them the use of the metals, and eventually displaced them from the fertile plains and valleys of Southern Europe. It seems reasonable to suppose that the Iberian tribes and the savage Ligurians, subjugated by the Romans, and described by Cæsar as dwelling in caves, may have been the southern representatives of the primitive folk, while the Finns and Lapps, as Nilsson suggests, may be the more modern and more northern representatives of the same folk, retreating northward with the retreat of the glacial fauna which followed the retreat of the glaciers. From the northern shores of Europe and Asia the same folk crossed to America; and the Esquimaux and North American Indians are the Stone folk in America, still following the pursuits of their ancestors—still using the bow, the kyak or canoe, and the stone hatchet, and perpetuating the Age of Stone in a remote land.

Primeval man, it must be admitted, was a barbarian, but he was by no means the stepping-stone between the apes and modern man. There is not a particle of evidence that he was not possessed of the faculty of speech, and did not exercise the same intellectual and moral powers as the citizen of the United States. Few human crania or other bones have ever been discovered upon which the judgment of the comparative anatomist could be brought to bear. Considerable diversity appears; but the skulls belong to the brachycephalic (or round-head) type, which, according to respectable ethnologists, was the type of the ancient Ligurian head.

Primeval man used the spear and the bow in his conflicts with the tiger, the bear, and the hyena, and in the wars which he waged with his fellow-man; he chased the elephant, the goat, and the musk-ox over the plains of Southern Europe, and fished with single and double pointed barbed hooks in the cool streams of Scandinavia. That he dwelt in caves we know. These were Nature's provision for the houseless. But there is no reason for supposing that he did not soon devise more comfortable dwellings. He seems to have resided at times upon the banks of rivers and by the ocean's shore. Whole villages, it would seem, must have cast into one common pile the refuse of their tables. These accumulations are sometimes several hundred yards in length, and from three to nine feet in height. The flint folk, whose household ware is mingled with the kitchen rubbish, must have dwelt in huts above the ground. At a somewhat later epoch we know that they drove piles in the lakes of Central Europe, and constructed platforms on which their dwellings were built. From these habitations they cast into the lake the refuse of their houses. By dredging, we recover stores of broken pottery, and implements of stone for cutting and for skinning, together with the bones of quadrupeds known to inhabit Europe in the Age of Stone. The dolmens of the same epoch prove also that primeval man understood the art of rough masonry.

There is no decisive proof that the earliest flint folk engaged in the cultivation of the soil or the domestication of the wild beasts. It is true that we find associated with human relics the remains of the hog, the dog, the ox, the horse, the sheep, the goat, the deer, the reindeer, the elephant, all of which have been domesticated in subsequent ages; and we certainly are not precluded from the presumption that some of these animals began to yield willing obedience to man even in this twilight epoch. We must cheerfully admit that these primitive people may have accomplished—undoubtedly did accomplish—many achievements of skill and intelligence of which it is now impossible to discover the record. Their food, like their dwellings, was at first supplied spontaneously by Nature; but during the epoch of the pile habitations, man seems to have learned the art of producing grain and vegetables. In some of the earthen pots dredged from the Swiss lakes have been found winter stores of fruits and cereals. Among them were beautiful specimens of wheat, and, in addition, barley, oats, peas, lentils, and acorns. At this epoch the people must have cultivated the ground and raised cattle. The discovery of millstones, with pestles of granite and freestone, shows that they knew how to grind their grain. The use of fire was known, and upon this they roasted their meat. They ate the marrow and brains of the animals killed, as we find the bones split open for the removal of these substances.

The clothing of the primeval folk was probably at first formed from the skins of quadrupeds; but during the age of the lacustrine cities they had learned the art of manufacturing textile fabrics, since among the other *débris* of the pile habitations have been found fragments of linen cloth. The garments, formed either of skins, bark, or cloth, were sewed together with needles and awls, of which the lacustrine cities furnish examples.

The man of this period was possessed of some degree of taste. This is shown first in the workmanship displayed upon the bone and horn handles of many of his tools, in the finish of lance and arrow heads, knives and daggers, in the fashion of his pottery, and in beads formed of pebbles, pieces of coral, and the teeth of wild animals. In some instances whistles have been found made of the digital bones of certain ruminants. His taste, and even no mean degree of artistic skill, are also displayed in the sculpturing of his tools and implements, and his delineations upon pieces of ivory, horn, and slate. "The decorations on many pots and implements," says Vogt, "consisting of simple, straight, angular, or crossed lines, exhibit a certain sense of beauty; but the drawings of animals, as discovered by MM. Lartet and Garrigou, are still more surprising. They are mostly engraved on bones, but also on slate. Those found by M. Garrigou represent heads and tails of fishes; those in possession of M. Lartet represent large mammals, among which the reindeer is easily recognized by the antlers. * * The masterpiece in Lartet's collection is a handle carved from the antlers of a reindeer, a real sculptured work, the body of the animal being so turned and twisted that it forms a handle for a boy's hand. All other drawings are in sharp and firm outlines, graved upon the surface of the bone, and it may be seen that the artist, in working it, turned the bone in various directions." The most interesting of all these relics of primeval art is the delineation upon ivory of the outlines of the hairy mammoth in a style which, though rudely and carelessly executed, leaves no doubt of the identity of the original of the picture. These people evidently possessed a marked predisposition to art. The rude hunter, wearied in the chase, amused himself in reproducing upon ivory and stone the forms that had excited his interest, and upon which undoubtedly he depended for subsistence and perhaps for service.

Lastly, primeval man was endowed with a religious nature. He formed numerous utensils consecrated to the ceremonies of religion. He buried his dead in grottoes closed with slabs, as the Jews continued to do at a later day. The recumbent position of many of the skeletons shows that, like the dead of the ancient Peruvians, they were entombed with an observance of religious rites. Like the American Indian, he provided his deceased friend with food and arms to supply his necessities while on the journey to another world. These are facts of extreme significance as tending to show that the religious consciousness, universal in our day, was also an endowment of the earliest and most uninstructed type of man.

The man of the Stone Age was not, therefore, as some have asserted, a sort of perfected monkey. He had the structure of a man; without doubt, he was capable of speech; he supplied his wants with a kind of skill which became improved and educated by experience—a characteristic only of intelligence; he admired beauty; he manifested a perception of the ideal; his thoughts strayed forward into another world, and, with his other religious sentiments, he undoubtedly felt a vague, strange sense of a superintending Intelligence and a moral Governor.

Does the unwritten history of this race reach back to an antiquity incompatible with prevalent views upon the age of man? Here, as elsewhere, the enemies of Revelation have sought materials for the use of unbelievers. They have sought in vain. There is more in the history of primeval man that confirms our Scriptures than there is of conflict with them. We have popularly held the race to be about six thousand years old; but our researches show that man lived with the bear, hyena, mammoth, and other animals now extinct, and some of which became extinct on the decline of the glacial epoch. It is not claimed that man lived before the glacial epoch, and the evidences of his contemporaneous existence with the reign of ice have been shown to be fallacious. The remains of man reputed to have been found in glacial drift of the valley of the Somme are in truth buried in deposits of much later date, as has been shown by Dr. Andrews, of Chicago, as well as by others.* Man had no place till after the reign of ice.

* For an elaborate and accessible paper on the "Amiens gravel," by Alfred Tyler, see Amer. Journ. of Science and Art [2], Nov., 1868, cited from the Quart. Jour. of the Geolog. Society of London for May, 1867.

But it has been imagined that the close of the reign of ice dates back perhaps a hundred thousand years. There is no evidence of this. The cone of drift materials accumulated at the mouth of the Tinière, in which have been found human remains, was estimated by Morlot to be from ninety-six to one hundred and forty-three thousand years old; but Dr. Andrews has exposed a curious arithmetical blunder, the correction of which reduces the time to within five thousand years. We have no rule for the measurement of post-tertiary time which necessitates the admission of so high antiquity to our race. If we have been accustomed to think of the extinction of the cave-bear as dating back to high antiquity, we now discover that he lived with man, and the reindeer, and other animals which still survive. The existence of even the cave-bear may not have been so very remote. What are the reasons assigned for the prevalent opinion that it was many ages ago that the glaciers began to disappear from Europe? Simply the existence at that time of quadrupeds now extinct, together with the presumption, unsupported, as it seems, by the facts, that no animals have coexisted with man except those of the recent fauna. The fact is, that we come ourselves upon the earth in time to witness the retreat of the glaciers. They still linger in the valleys of the Alps, and along the northern shores of Europe and Asia, while the disappearance of animals once contemporaries of man is still continuing. Not only did contemporaries of man become extinct during the Age of Stone; some survived to the twelfth, fourteenth, and sixteenth centuries, as already stated; the Moa of New Zealand, and the Æpiornis of Madagascar, have become extinct within the epoch of tradition, as indeed has the Mammoth of North America; the Dodo of Mauritius disappeared in the seventeenth century; the Great Auk of the arctic regions has not been seen for half a century;

and every one must be convinced that the beaver, elk, panther, buffalo, and other quadrupeds of North America are approaching extinction by perceptible steps. The fact is, we are not so far out of the dust, and chaos, and barbarism of antiquity as we had supposed. The very beginnings of our race are still almost in sight. Geological events which, from the force of habit in considering geological events, we had imagined to be located far back in the history of things, are found to have transpired at our very doors. Our own race has witnessed the dissolution of those continental glaciers which we have so long talked of as incidents of pre-Adamic history. Our own race has witnessed the submergence of Southern Europe; the detachment of the British Islands and Scandinavia from the continent; the wanderings of the great rivers of Eastern Asia; the submergence of thousands of square miles of the coast of China, so that the seats of ancient capitals are now rocky islets far at sea; the emergence of the ancient country of Lectonia; the drainage of the vast lake which once overspread the prairies of Illinois; the alternations of forests, and many other events which we once associated with high antiquity. It is the opinion of Hooker and Gray that the Falkland Islands, and others in the vicinity, have formed a part of the continent of South America during recent times, and that during this connection they acquired the continental fauna and flora. The Straits of Behring may even have been cut through since the early migrations of man and his contemporaries, the mammoth and reindeer; as in some distant future age the Isthmus of Darien, which now connects North and South America, may become a strait separating them. There is no more reason in this day than fifty years ago to claim a hundred thousand years for the past duration of our race.

I can not refrain from noting the peculiar relief which

the mind experiences in discovering the means to seize and comprehend some of the oppressively vast cycles which geology discloses. Here is a geological age—the Post-Tertiary Age—unfinished, it is true—which we almost possess the means of measuring. The life of our race reaches back beyond grand geological events. We have some notion, from the progress which our race has made during the period of written history, what must have been the duration of its infantile tutelage. Nay, the records of the Somme and the Tinière, as we now decipher them, afford us a common measure of the age of man and the duration of the Post-Tertiary. The vast changes that have transpired upon the coast of China, the shores of the Mediterranean, and other parts of the world, since man has been a beholder of geological history, seem to carry us back into the midst of the grand events which we have so solemnly and wonderingly contemplated from our seeming distance. These geological intervals, after all, are appreciably finite. The discovery affords a sensible relief to the mind so long oppressed by the contemplation of cycles which lose themselves in the haze of eternity.

One farther thought crowds itself into the company of these reflections. It is a thought of the growing perfection and exaltation of our race. How have we struggled through many ages, upward from companionship with beasts, from clothing of skins or bark, houses of caves, implements of chips of flint, a vague consciousness of a Superior Being—like the polyps' sense of light felt through all its body—through all the grades of pupilage, all the degrees of civilization, all the heights of mental and moral exaltation up to man as he now is! What a picture of progress is here! How abject once—how exalted, how spiritualized, how God-like now! Is not man approaching nearer to God? How vastly less of the brute—how infinitely more of the spiritual! Once he contented himself to capture prey sufficient for food, as the bear and the tiger did in whose company he lived. But—oh, how unconscious of his powers! he held even then the spark of divinity which the bear and the tiger had not, and he has risen, while they grovel on the plane from which he sprang. From age to age he has learned to commune more and more with the unseen—the ideal—the good and the true. He has made achievements which were once beyond the reach of dreams. Steam, electricity—what miracles do they not summon into mind? What does a retrospect of fifty years disclose? And is not man even yet on the march of improvement? What does a forward glance of fifty years unfold to imagination? What now irresolvable mysteries may not be explained in the school-books of our grandchildren? There is nothing which it is reverent to pronounce inscrutable among the works of God. It remains for us to penetrate the world of invisible things. We have already sundry rumors and pretences—shadows cast before, perhaps—but as yet unsatisfactory and unintelligible, and, above all, unreduced to a philosophy. There must be a substratum that has not yet been sounded lying beneath the confused and apparently capricious phenomena of clairvoyance, mesmerism, dreams, and spiritual manifestations. With much imposition, there is much which can not be scientifically ignored. It remains to resolve the mystery of these sporadic phenomena—to reduce them to law, and to open under the law some regular and intelligible intercourse with the unseen world. The unseen world is destined to become like a newly discovered continent. We shall visit it—we shall hold communion with it—we shall wonder how so many thousand years could have passed without our being introduced to it. We shall learn of other modes of existence—intermediate, perhaps, between body and spirit—having the forms and limitation in space peculiar to matter, with the penetrability and invisibility of spirit. And who can say that we may not yet obtain such knowledge of the modes of existence of other bodies as to discover the means of rendering them visible to our bodily eyes, as we now hold conversation with a friend upon the shores of the Pacific or in the heart of Europe, or fly with the superhuman velocity of the wind from the Atlantic to the Mississippi Valley. Then may we not at last gaze upon the "spiritual bodies" in which our departed friends reside, and discover the means of listening to their spirit voices, and join hands consciously with the heavenly host? Oh, who can say what these exhaustless and illimitable powers of the noble soul of man may not accomplish? Does the reader smile? I believe these are the suggestions more of philosophy than of fancy. Does he say it is only a dream of impossibilities? He assumes that he knows every thing which the infinite Intelligence can fathom. To fetter the human soul with assumed impossibilities is impiety. The bird which would soar first looks upward. The soul never attains that which it does not strive for. If we would commune consciously with the unseen world we must have both faith and works. In reference to the perfectibility and exaltation of the intellectual and moral nature of man, let no one say "impossible."

CHAPTER XXXIII.

WILL THERE BE AN ANIMAL SUPERIOR TO MAN?

HE that has glanced over the long line of organic history, and observed how the ascent from the sea-weed to man has been effected, step by step, in regular succession, can not fail to start the inquiry, "Is man destined to be the last term of this series of improving types?" I reply that, while this is peculiarly a question to be answered by Revelation, science affords some intimations which tend to assure us in the possession of the dignity which we now enjoy as the archonts of terrestrial existence.

In the first place, all geological preparations and ideas converge in man. The world seems to have been designed with the view of stimulating to activity the powers of a thinking being. The universe is a rational product; and every department of it, and every isolated object, sustains an intelligible relation to other parts and objects. We are not left to infer, or even to know, that intelligent design is locked up in the secret plans of creation; but what is more suggestive, as well as more satisfactory, is the fact that this intelligence is patent before our eyes, so that we read, as it were, a revelation of the thought embodied in the works of the visible universe. And much of that which is not at once manifest yields to investigation, while a stimulus to investigation is found in the hints and suggestions which Nature seems intentionally to have dropped along the pathway of him who follows the beckoning of his thoughts. Not only were these germs of thought planted from time to time during the whole progress of the past creation; and not only is man the first creature capable of responding to the stimuli to mental activity, but more; this mentality, while it differs qualitatively from the highest endowments of the lower animals, is in itself the highest possible grade of endowments. It is qualitatively identical with that infinite Intelligence whose presence and supremacy are recognized throughout the universe. It is a fair presumption that when the course of animalization has attained the point toward which all these intellectual adaptations converge, a point is reached which will not be passed except under a different general scheme.

Similar remarks apply to the co-ordination existing between the material world and the idea of the beautiful in man. The beauty and sublimity of Nature have no relation to any other creature. Man is the consummation of a dualism. While the beautiful implies man, it excludes his successor. No endowment beyond or higher than a response to the provisions of Nature is possible.

The beneficent provisions of the earth's crust not only prophesy man, but they reach their finality in man. It was only for human uses that the coal was treasured in the recesses of the earth; for human uses alone the mountains have lifted up their burdens of iron; for human uses only the grandest movements of geological history elaborated and distributed a soil. It is only for man that the forests yield their abundant supplies of timber and fuel. For man the edible and medicinal vegetables were provided. For man the natures of the domestic animals were moulded; and their domestic attachments are directed to no other being.

The last geological revolution produced results of a general rather than a local character. During the Paleozoic, Mesozoic, and earlier Cenozoic ages, the action of geological agencies had been especially developed along belts parallel to the main bodies of land. In the glacial epoch, however, a phenomenon occurred which, so far as we know, was unprecedented in its universality. The whole northern portion of both continents was covered by glaciers, whose effects were felt in America to the Ohio River, and whose *débris* were borne, in the next epoch, to the Gulf of Mexico. This sudden extension of the range of geological activity was something paralleled by the release of the human species from those restraints which confined all preceding animals within narrow limits, and constituted, like that, an indication that a full pause had been reached in continental preparations—as when the sculptor, after having developed singly, with time and care, the individual features of his work, subjects it finally to that general treatment which imparts the smooth and finished surface.

Lastly, it may be added that vertebrate development both points toward man and attains its consummation in man. The earliest fish which moved in the waters of the Paleozoic seas embodied, in its osteological organization, a prophecy of man; the Mesozoic reptile still pointed onward toward man; the Tertiary monkeys were a higher summit of vertebrate organization from which the yet higher Alp of human structure was still pointed to, illumined by the rising dawn of the modern world. In the skeleton of man we have, at last, the fulfillment of the prophecies of ages.

Man stands in the focus of all the conceptions embodied in past history. We are as little authorized to allow that the course of development is destined to advance beyond him, as to deny that it has furnished intimations, in all ages, that it was destined to reach to him.

Consider, in the second place, man's superiority over the brutes. Among the myriads of animals which populated the earth during the cycles of geological history, supremacy was the reward only of superior force. Man gains supremacy through his intellect. Brutes dominate through the physical forces belonging to matter; man, through the immaterial forces which are the attributes of Deity.

The chasm which separates the intelligence of man from that of the brutes is broad. It is not simply a step in the easy gradations observed among the brutes themselves; it is a break in the chain of gradations. Even if not qualitatively superior to that of brutes, its sudden expansion is so great that its sphere of activity creates a new quality in the being. Man is the first being in all the history of the world that could contemplate creation, and abstract the intelligence displayed in it, and experience a glow of satisfaction in attaining to the thoughts first conceived in the mind of the Omniscient. Man is the first animal capable of contemplating Deity. In these exalted endowments not only does he excel the brutes, but he excels them in so vast a degree as to suggest the belief that the gradations of animal existence had been concluded, and Nature had reached a full pause. The material part—the frame-work—of animality had been perfected by slow gradations; and now, on the creation of man, Nature superadded an unprecedented endowment—a spiritual organization which makes man both a prince and a masterpiece in creation.

When we speak of man's moral nature we touch a subject which recalls all that has just been said of his intellect, and affirms it with redoubled emphasis. There are reasons for believing that this endowment differs in kind from any thing in the nature of the brute. This, to the ability to understand God, adds the ability to sympathize in his moral attributes, and to enter into moral relations with him and with humanity. Man stands in contact with God. A farther approximation is impossible. He must be the limit, as he is the existing culmination of organic life.

These various considerations, with others, seem to teach

that the column of organic succession is complete in man. The lower forms, gradually and regularly ascending from base to summit, constitute the shaft of the column; but in man we have a sudden expansion, an ornateness of finish, an incorporation of new ideas, which designate him as the capital and completion of the grand column of organic existence.

Consider, in the third place, man's unlimited geographical range. When the first animals were introduced upon the earth, they found the ocean encompassing it on every side, and creating a uniformity of physical conditions which enabled them to range through every latitude and longitude. In later ages, as the continents, with their mountain ranges, became differentiated from the terrestrial mass, and diverse climates were called into existence, we find that animals were restricted to successively narrower limits. Not only did the growing differentiation of the different regions of the earth lead toward the restriction of the faunas, but there is something in higher organisms themselves which specializes them in their adaptations, and unfits them for so wide a range, even with external conditions unchanged. Thus, as animal life advanced upward, it became more narrowed in the range of its species. The species in possession of the earth immediately previous to man were more restricted than any of their predecessors. It would certainly be expected from all these analogies that man, on his appearance, would be limited to the narrowest bounds of all. What is the fact? Man overleaps all barriers. Climates, mountains, oceans, deserts, form no impediments to his migration. He, the first of all animals, has literally extended over the whole earth, and fulfilled the command to take possession, to use, and to enjoy. What does this signify, if not that man is the completion of the series? Animal existence, first narrowed to the smallest limits in its specific range, then suddenly expanded to the widest. Man occupies the whole earth; he is not only the finishing stroke, but he excludes a successor.

Consider, lastly, man's erect attitude. When the fish, the earliest representative of the type which embraces man, was introduced into the waters of the Devonian seas, the vertebral axis was hung in a horizontal position, and the animal was not endowed with even the power to raise the head by bending the neck. Many of the Carboniferous fishes acquired this power, but they remained suspended in the element of lowest vital relations. The Triassic and Jurassic Enaliosaurs, while they continued to inhabit the water, breathed the air, and held the head habitually a little elevated. The Crocodilians to these endowments added the power to *crawl* upon the ground. The Deinosaurs of the Cretaceous Age *walked* upon the land with the body elevated above the ground, but the head remaining nearly horizontal. The birds assumed an oblique position of the spinal axis; and most of the Tertiary mammals, which followed them, could carry their attitudes from the horizontal to the semi-erect. The higher monkeys lived normally in a sub-erect position, but still supporting themselves by the four extremities. Man first and alone assumed a perpendicular attitude, and turned his countenance toward heaven, and talked with the Being who formed him.

> "Prona cum spectent animalia cætera terram,
> Os homini sublime dedit; cœlumque tueri
> Jussit, et erectos ad sidera tollere vultus."

It is evident no farther progress can be made in this direction. The elevation of the spinal axis has reached a mathematical limit; the consummation of organic exaltation is attained.

These various considerations concur in justifying the assumption that the Author of Nature regards his work as completed. The universal belief of the Christian world, therefore, that the termination of the existence of the human race will mark the consummation of the history of the present order of things, seems to be founded equally in our mental constitution and in the philosophy of the material creation.

CHAPTER XXXIV.

POPULAR BELIEFS IN PERIODICAL CATASTROPHES TO THE UNIVERSE.

WHENCE come we, and whither are we tending? Whence this ponderous globe which we inhabit? What vicissitudes has it undergone? What is its final destination? And when the drama of the world is closed, what then? Whence this magnificent system of a visible universe? and of what inscrutable purposes does it form a part? What is that which is first of all—the cause of all—self-existent, uncreated, without beginning and without end?

These are grand problems—the most stupendous with which the human mind can grapple. We can not presume to offer their final solution, but we may venture to inquire what light is thrown upon their solution by the converging rays of all the sciences.

These are problems which have engaged the attention of thoughtful minds in every age of the world. If we look into the pages of ancient philosophy, we find it every where occupied with inquiries into the origin and destiny of the universe—the different orders or kinds of existence—the absolute existence, on which all other being depends—the nature of Deity and of man, and their relations to each other and to other grades of existence. These have been the great, ever-present, obtrusive mysteries with which the human mind has always been grappling. On the shores of classic Greece we find Thales, Pythagoras, Zeno, Epicurus, Plato, and a long and brilliant line of thinkers pondering over the problems of mind and matter. On the other side of the Mediterranean we hear the same interrogatories resounding from the region of civilization's dawn, in Egypt, and in far-off India and China other races have found themselves confronted by the self-same mysteries, and, with equal courage, have demanded from the depths of Nature their solution. These sublime questions have stared with equal steadiness in the face of Greek, Egyptian, Phœnician, Chaldæan, Jew, Persian, Arabian, and Hindoo. Perennial problems, omnipresent as mind itself, they have reappeared upon American shores; and we find that the sacred books of the Aztecs yield us a cosmogony and a theogony no less sublime than those of India, Persia, and Greece.

Problems which, in all ages, have stood foremost in the conflict of the human mind with the vast unknown, would mock at the attempt to grapple with them in the brief compass of a chapter or two; but we can not pass them by without taking a few bearings upon their salient points. Waiving entirely the questions which arise in reference to moral and intelligent existences, let us attempt to bring together a body of considerations bearing upon the doctrine of periodical destructions and renovations in the material universe. It will thus, I think, be made to appear that the existing order of things *is not eternal*, and that a crisis is approaching which will demand the interposition of *a power superior to Nature.*

Dr. Reid, the Scottish metaphysician, asserts that God has implanted in the mind of man an original principle by which he believes in and expects the continuance of the course of Nature. This, evidently, is an error, since our expectation of the continued recurrence of natural phenomena in the same order is based upon our past experience, and is, consequently, an induction instead of a necessary truth. The fact is, that in all ages of the world, and among every people who have attained to a philosophic system, the contrary belief has been prevalent. The existing order of Nature has been regarded as temporary, and the flow of terrestrial and even of cosmical events has been conceived as destined to be broken up by universal revolutions.

The Chaldeans, according to Berosus, held that the world is periodically destroyed by deluges and conflagrations. The deluges they believed to result from a great conjunction of the planets in the constellation Capricorn, and the conflagrations from a similar conjunction in the constellation Cancer. Some of the Christian fathers adopted these views. The Chaldeans also calculated the end of the world from the period occupied in the retrograde movement of the stars through one complete circumference—a phenomenon due to the precession of the equinoxes, and accomplished, as modern science has shown, in a period of 21,000 to 26,000 years.

The Chaldean philosophers had also their Annus Magnus, or Great Year, at the end of which the present terrestrial and cosmical order would be brought to a termination by an ordeal of fire, after which it would be again renewed.

The ancient Scythians, in their dispute with the Egyptians in reference to the relative antiquity of their respective nations, reminded them that the world undergoes revolutions both by fire and water.

The Egyptians, according to Plato, fancied that the heavens and earth originated in a promiscuous pulp. From this the elements separated of their own accord; fire sprang from the upper regions; the air began to move. The warmth of the sun bred living creatures innumerable in the plastic mud, and these, according to the predominance of the various elements, betook themselves to the air, the water, or the solid land. Man was generated from the slime of the river Nile. By a gradual improvement of the lower creatures, and a gradual perfection of the globe, the world became what the Egyptian found it, and was destined to flourish through an interval of time expressed by their Annus Magnus, or great year—a cycle composed (as with the Chaldeans) of the revolutions of the sun, moon, and planets, and terminating when these return together to the same sign whence they were supposed to have set out. The duration of this great cycle, according to Orpheus, was one hundred and twenty thousand years; according to others it was three hundred thousand; and by Cassander it was taken at three hundred and sixty thousand years. At the end of each great year or cycle the world was supposed to be subjected to the destructive ordeal of fire or water, by which it was renovated, to become the abode of a regenerated race of men.

The Hindoo cosmogony, which was perhaps the germ of all that was taught by the Western nations, gives prominence to the doctrine of secular catastrophes and renovations. "The First Sole Cause," say the Institutes of Menu, "thinks within himself, 'I will create worlds.'" Water is then brought into being, over the surface of which moves Brahma, the Creator. Brahma first effects the emergence of the land from the waters, and the creation of the firmament. He then vivifies the earth, in succession, with plants, animals, celestial creatures, and man. The sun springs from his eye, the air from his ear, the fire from his mouth. From his mouth, his arm, his thigh, his foot, proceed the founders of the chief Hindoo castes. Brahma, having accomplished his task, "changes the time of energy for the hour of repose." He sleeps during 4320 millions of years—a day of Brahma—at the end of which time the world is destroyed by fire. The flames are at length quenched by the fall of incessant rains for a hundred years, and the waters, overspreading the earth, fill the middle region and inundate heaven. The world is enveloped in darkness, and the universe is reduced to one vast ocean. The breath of Vishnu next becomes a strong wind by which the clouds are dispersed, and the Deity then appears in the form of Brahma reposing on his serpent couch upon the deep. As soon as he awakes the world is renewed, to be again destroyed and again renovated after each kalpa, or day cf Brahma's existence. "For there are creations and destructions of worlds innumerable." At the end, however, of a hundred years, each consisting of 360 kalpas, and each kalpa of 4320 millions of our years, Brahma himself, and all things with him, will cease to exist.

Among the Jews there has been extant, from time immemorial, a prophecy that the world was destined to endure 6000 years—2000 before the Flood, 2000 under the Law, and 2000 under the Messiah. This belief is cordially accepted and strongly insisted upon by a majority of the Christian fath

From the East the doctrine of periodical revolutions found its way, with the migrations of men, into Europe. The Persians, the Chaldeans, the Egyptians, and the Phœnicians adopted it in Western Asia and in Africa, while the "Orphic Hymns" afford us the earliest germination of the Eastern faith in Greece. Orpheus and Menander, who flourished in the very twilight of Greek poetry and civilization, and who undoubtedly derived their philosophy from the Egyptians, reproduce the myth of the Annus Magnus, and teach that the universe is destined to be dissolved on the completion of this cycle. Like the Indians and Jews, the authors of the Orphic Hymns assigned a definite duration to the Annus Magnus, as has been already stated.

In the Sibylline Books, whose origin dates back, perhaps, 1300 years before our era, this ancient faith is shadowed forth in another guise. The world is destined to endure ten ages, the first of which is the Golden Age. After a renovation by fire the Golden Age will return, when, according to the authority of Virgil, the serpent will perish; the earth will produce her crops spontaneously; the kid will no longer fear the lion; the grape will be borne upon the thorn-bush, and scarlet, and yellow, and royal purple will become the native colors of the woolly fleece.

> "Ipsæ lacte domum referent distenta capellæ
> Ubera; nec magnos metuent armenta leones.
> Ipsa tibi blandos fundent cunabula flores;
> Occidet et serpens, et fallax herba veneni
> Occidet; Assyrium vulgo nascetur amomum.
> * * * * *
> Molli paulatim flavescet campus arista,
> Incultisque rubens pendebit sentibus uva,
> Et duræ quercus sudabunt roscida mella."

The Stoics, who derived the doctrine from the Phœnicians, and were its principal advocates among the Greeks, held that the world would be destroyed by a conflagration. This they thought would occur "when the sun and stars shall have drunk up the sea." "Yes," says quaint old Thomas Burnet, "but how long shall they be a drinking it?" The Stoics, in speaking of the restoration of the earth after the final conflagration, employ the same terms as we find in the sacred Scriptures. This, to say the least, is an interesting coincidence. Chrysippus calls it "Apocatastasis"—restitution—as St. Peter does in the Acts. Marcus Antoninus several times calls it "Palingenesia"—regeneration—as our Savior does in Matthew, and Paul in the epistle to Titus; and Numenius has the two scripture terms "resurrection" and "restitution."

The doctrines of the Pythagoreans—save a few, who in later times were led off by Aristotle—were nearly identical, in respect to periodical revolutions, with those of the Stoics. Like the philosophy of the Stoics, that of Pythagoras was drawn from the older civilization of Egypt and Persia.

The Ionics discoursed much of the origin of things, and agreed with the Epicureans and Stoics in their doctrine of secular catastrophes.

Plato, the preceptor of the Academics, admits that the earth is subject to the transformation of deluges and conflagrations, but expresses the belief that the universe, as a whole, is something so beautiful and noble that the goodness of God will perpetuate its existence.

Aristotle alone, of all the ancient philosophers, maintained the eternity both of the matter of the universe and of the existing order. He confesses to a pride in this, since the doctrine, as he claims, is at variance with the unanimous belief of antiquity.

Among the Romans, Lucretius, Lucan, and Ovid openly discourse upon the prevalent doctrine of periodical catastrophes; and Cicero, who intermeddled with all learning, assures us that the memory of mighty deeds can not be eternal, since conflagrations and deluges periodically obliterate all record of human achievements.

The Celts, according to Strabo, held the same traditions in the west as were current among the nations of the east of Europe. Their Druids secured the world an immortality only through periodic ordeals by fire and water.

The Persians represent their god, Fire, as the final avenger of the sins of men, and the destroyer of the world.

Among the Arabians and Indians, the story of the Phœnix is an allegory of the earth. This bird of fable no sooner crumbles to ashes than she rises again in more than pristine beauty. They have a similar fable of the eagle, which is represented as soaring so near the sun as to renew his youth. Allusion seems to be made to this myth in the Psalms, where David says: "Thy youth is renewed like the eagle's"—a passage which the Chaldee paraphrase renders, "Thou shalt renew thy youth like the eagle in the world to come."*

The Aztecs, according to Humboldt, felt the curiosity common to man in every stage of civilization, to lift the veil which covers the mysterious past and the more awful future. They sought relief, like the nations of the old continent, from the oppressive idea of eternity by breaking it up into distinct periods or cycles of time, each of several thousand years duration. There were four of these cycles, and, at the end of each, by the agency of one of the elements, the human family was swept from the earth, and the sun blotted from the heavens, to be again rekindled. The Aztec's conception of the origin of man is nobler, and more approximated to that of the Jewish Scriptures, than either the Egyptian or the Hindoo. The following are extracts from a translation of the Popol Vuh, or National Book of the Quiches of Guatemala. How marvelously conformable is the first extract to the story of the earth as recited by geology!

"There was not yet a single man; not an animal; neither birds, nor fishes, nor crabs, nor wood, nor stone, nor ravines, nor herbs, nor forests; only the sky existed. The face of the land was not seen; there was only the silent sea and the sky. There was not yet a body, naught to attach itself to another; naught that balanced itself, naught that made a sound in the sky. There was nothing that stood upright; naught there was but the peaceful sea—the sea silent and solitary in its limits; for there was nothing that was. * * * Those who fecundate, those who give being, are upon the waters like a growing light. * * * While they consulted the day broke, and at the moment of dawn man appeared. * * * Thus they consulted while the earth

* "In mundo venturo renovabis, sicut aquilæ, juventutem tuam."

grew. Thus, verily, took place the creation as the earth came into being. 'Earth,' said they; and the earth existed. Like a fog, like a cloud was its formation; as huge fishes rise in the water, so rose the mountains; and in a moment the high mountains existed."

The foregoing extract is from the history of the first creation. It can not be necessary to point out the parallels between this passage and the pictures drawn by the classic poets—especially Ovid—nor even to direct attention to the points of coincidence with the Mosaic account of chaos and incipient order. The following passage is from the account of the fourth and last creation:

"Hear, now, when it was first thought of man, and of what man should be formed. At that time spake he who gives life, and he who gives form, the Maker and Moulder, named Tepen, Gucumatz: 'The day draws near; the work is done; the supporter, the servant is ennobled; he is the son of light, the child of whiteness; man is honored; the race of man is on the earth;' so they spake. * * * Immediately they began to speak of making our first mother and our father. Only of yellow corn and of white corn were their flesh, and the substance of the arms and legs of man. They were called simply *beings*, formed and fashioned; they had neither mother nor father; we call them simply *men*. Woman did not bring them forth, nor were they born of the Builder and Moulder, of Him who fecundates and of Him who gives being. But it was a miracle, an enchantment worked by the Maker and Moulder, by Him who fecundates and Him who gives being.

"Thought was in them; they saw; they looked around; their vision took in all things; they perceived the world; they cast their eyes from the sky to the earth."

"Then they were asked by the Builder and Moulder, 'What think ye of your being? See ye not? Understand ye not? Your language, your limbs, are they not good? Look around beneath the heavens; see ye not the mountains and the plains?'

"Then they looked, and saw all that there was beneath the heavens. And they gave thanks to the Maker and the Moulder, saying, 'Truly, twice and three times, thanks! We have being; we have been given a mouth, a face; we speak, we understand, we think, we walk, we feel, and we know that which is far and that which is near. All great things and small on the earth and in the sky do we see. Thanks to thee, O Maker, O Moulder, that we have been created, that we have our being, O our Grandmother, O our Grandfather!'"*

I can not help regarding these sentiments—these reveries of the uninspired and uninstructed intellect of man feeling after the mystery of his origin and the origin of created things—as equaling in sublimity the contemplations of a Socrates or a Plato groping by the dim light of reason for an outlook into the future of the soul.

* Histoire des nations civilisées du Mexique et de l'Amerique centrale, durant les siècles antérieurs à Christophe Colomb, écrite sur des documents originaux et entièrement inédits, puisés aux anciennes archives des indigènes, par M. l'abbé *Brasseur de Bourbourg*. 4 forts vol. in -8 raisin avec carte et figures.

CHAPTER XXXV.

SOME THOUGHTS ON PERPETUAL MOTION.

FROM the citations made in the last chapter we discover the existence of a unanimity of belief in the doctrine of periodical catastrophes which is well calculated to excite a spirit of scientific curiosity. It can scarcely be attributed to a mere tradition descending through the ages, and through all the nations between us and the ancient sages upon the banks of the Ganges. Mere tradition is generally circumscribed by the nationality or race among whom it originates. A tradition of a philosophic character must have been subjected to the scrutiny of the philosophers of the nations to which it traveled. If admitted, and maintained, and perpetuated from age to age among different nations, it must have been because recognized as something more than a tradition. The philosophy of Greece and Rome never harbored a tenet which could only be defended as an Oriental tradition. It must have discovered some rational grounds for the acceptance of this belief, and thus have made it a philosophic principle.

What were the grounds of the naturalization of this Oriental faith we might be unable to determine. Pythagoras, however, explicitly taught that his faith was founded on an observation of geological phenomena; and Lyell thinks that the doctrine in general was based upon records and traditions of deluges and earthquakes, any of which came far short of revolutionizing the face of the earth.

A doctrine so ineradicable, and so spontaneous in every soil, must have rested upon a *rational* belief. That belief may be of the nature and authority of an intuitive sentiment. The unanimous consent of mankind to any proposition is to be regarded as the utterance of humanity. That which our common humanity expresses is the expression of the Author of our humanity; it is a kind of revelation, and will be found in all cases to correspond to a reality.

But we are not compelled to refer this doctrine to any spontaneous, and universal, and necessary intimations growing out of the constitution of human nature. Why may not this faith have been a grand generalization reached in common by the philosophic minds of all ages? The facts of Nature have always been patent to all the world. The phenomena upon which we have reared the stupendous structure of the modern sciences were as open to the scrutiny of Thales, and Pythagoras, and Plato as to us. There are *scientific grounds* for such beliefs; and the ancient sages, though they certainly failed to appreciate the data of science to the same extent as ourselves, may reasonably be supposed to have caught glimpses of majestic inductions which involved the destruction of terrestrial order, or even the order of the material universe.

We stand now in the presence of those grand and instructive phenomena. On an eminence in the midst of the visible universe, with the multitudinous events of earth and heaven transpiring before our eyes—a universe flooded by the ethereal light of modern science—our intelligence gifted with the power to penetrate to the core of the earth, or fly beyond the flight of the most erratic comet—or pierce the gloom of a million ages passed—or lift the veil which opens the vista of a million ages to come—and here, in this favored position, we ask ourselves what tides we witness in the flow of terrestrial and cosmical events. It is a sublime query. With boldness, but with humility and reverence, let us seek the answer.

Looking around us, we behold all Nature instinct with motion. The winter winds are hurrying to and fro; the storm-cloud scatters moisture from the mountains to the sea; the noisy torrent foams down the hill-side, and the majestic river winds ceaselessly to the ocean; vapors rise from the ground and descend again in rain and snow; the punctual tide performs unweariedly its daily perambulation of the globe; the waves' hoarse growl along the rocky beach is never stilled. The forces of matter, in their multiple forms and their myriad labors, keep every element and every atom constantly astir. If we look up, the sun, and moon, and stars are on their journeys. Every planetary orb and every satellite is full of motion. Even while it performs its stupendous journey about the sun, it is forever shifting its attitude in respect to itself. Not content with orbital and axial motions, each planet nods grandly from its ethereal altitude, and keeps time with the rhythm of the solar year. The stars which we call "fixed" are probably in motion, since twenty or thirty pairs of stars are seen to revolve about each other; and, if the wonderful induction of Mädler is to be credited, our sun, with his retinue of over a hundred planets, satellites, and comets, is sweeping through space on a stupendous journey of 18,000,000 of years.

Now we start the inquiry whether all this motion can be perpetuated forever. Motion, according to the new philosophy, is but one of the modes of heat, or electricity, or light, or magnetism, or chemical affinity. Under certain circumstances, one of these forms of force is changed into another. It is a law of every form of force *to seek a statical equilibrium*, and the transformation of a force signalizes its attainment of an equilibrium. A hammer descends upon a bar of steel and comes to rest; the *motion* is counteracted, but at this instant, and in consequence of its disappearance in the form of motion, it reappears in the form of *heat*. This heat seeks an equilibrium by transferring itself to the colder air, in which motion reappears in the heated ascending column. But this motion, in turn, disappears when the heated column, by transference of its heat, has ceased to be warmer than the contiguous air. All force is seeking some affinity with which it may be at rest, or it is striving to effect a motion which will bring its activities to rest. In obedience to the force of gravity, rain falls from the clouds, gathers itself into little rills, which, uniting their forces, join arms with the brooklet, and thence glide in company with the rivulet to the outlet of the valley, and wend their way to the sea. In the deep bed of the ocean the waters rest. The demand of gravity is satisfied. The friction of ascending vapors upon the atmosphere disturbs the equilibrium of the electricities; they flash in anger from cloud to cloud, and between the clouds and the earth, ever striving to restore the equilibrium. When that is effected, all the phenomena resulting from electrical action cease, and would forever cease were not a fresh disturbance introduced. If the electricities are again disturbed, it is because *some other force* is seeking its equilibrium. This other force is out of equilibrium because *some third force* has created disturbance in the search for its own equilibrium, and thus link hangs upon link in this chain of causation. We know not how far back the remotest disturbing force may lie, but of this we may be certain; there is somewhere, or will be somewhere in the future, a last disturbing force. Behind this, all is rest. When this has attained its equilibrium, all the phenomena resulting from the struggle of the forces will cease.

This is a mere abstract statement of the case. It possesses a higher significance than we may suspect. The argument concerns the stability of the very earth on which we tread. Every one has heard of the chimera of "perpetual motion." Not every one, however, has considered that the impossibility of perpetual motion results from the impossibility of transforming forces in a perpetual circle. Force shuns perpetual motion. It tolerates no such monotony. It is seeking rest. In larger or smaller quantities it steals away from you, and lies down to a quiet slumber, while your machine is deserted and motionless as a corpse. Heat filters in every direction through the atmosphere; motion steals through the bearings of your wheels, and, under the guise of frictional heat, it sneaks away from your control.

All motion is *mechanical.* There is no motion in the heavens above, or upon the earth beneath, which is not effected by the self-same forces as we incorporate in a steam-engine, or vainly strive to chain to the drudgery of perpetual motion. Every movement which we witness upon the earth—whether of winds, or clouds, or waters, or quaking mountains, is but the motion of *some part of a machine.* The earth is a piece of mechanism. The varied motions which we witness upon its surface arise from the perpetual transformations of force. The solar system is a piece of mechanism. All its visible motions have been demonstrated to arise from the action of the same force as that which drives a water-wheel or a hydraulic ram.

The question then arises whether the motions of a great machine are more likely to be perpetuated than those of a small one. A vast and complicated machine can be nothing more than a concatenation of small ones. The very statement of the case suggests a negative response. Terrestrial forces, like those which impel the locomotive, are wearing themselves out. All their activities are destined to be invaded by the sluggishness of age—by the torpor of death. The cosmical machine, like a clock, is running down, and, like a clock, will eventually demand the interposition of an Intelligent Will to re-establish its motion. The denial of this proposition drives us to one of the following alternatives: first, that there exists in Nature an endless series of causation—the remotest assignable cause still hanging upon another cause not higher than a material force—a conclusion entirely at variance with our intuitive cognition of primary causation; or, secondly, that one or more of the series of causes can act in different modes, so that what had just been done is presently undone, or done differently, and thus new conditions created for the renewed activity of other forces. But the supposition of a change in the mode of action of any force or cause contradicts a fundamental axiom of philosophy. We have no authority for such an assumption, and are not at liberty to resort to it.

It can not be denied that these are conclusions which are repugnant to the popular apprehension of Nature's operations. The thought of a "machine," moreover, suggests self-action, and seems at first to exclude that intelligent special agency in Nature which we call Providence. The solar system is, nevertheless, a combination of matter and force whose movements can be calculated with the same precision as those of a steam printing-press. If it be necessary to protect our notion of a Providence, let us suppose that those mighty forces which handle planets as if they were engaged in a "game of ball" are *not* energies inherent in matter, but *the immediate efforts of a divine will.* It may be so. There is no logic which can overthrow the assumption. But in either case, these energies are put forth according to intelligible and *unvarying methods;* and all that science asserts is, that *if the methods remain the same*—that is, if the laws of Nature continue unchanged—the course of cosmical activities will complete its round.

All the material forces, therefore, of the universe, both mechanical and physiological, with their actions and reactions, their equilibria and perturbations, are tending gradually toward a general and permanent rest. The threads of their mutual connection may be closely interwoven, but somewhere there is a beginning and an end. Within the grand cycle of their active lifetime apparent circles may be described, but, like the eddies of a river stream, they are lost in the general current, or, like the gyrations of a disk descending through the sea, they are only apparent, and wend their way toward ultimate rest. The same exact conditions are never reproduced. [See Appendix, Note IX.]

CHAPTER XXXVI.

WILL THE MOUNTAINS BE LEVELED?

LET us now direct our attention to a more specific examination of the circumstances under which the visible activities of our terrestrial abode are carried on. The fact which first and most strongly arrests our attention is the presence of universal and perpetual change. This fact alone demonstrates that the existing terrestrial order *had a beginning*. Work is in progress before our eyes; we may easily determine what has been accomplished and what remains to be accomplished. Had these changes been in progress from all eternity, every thing which existing forces are capable of effecting would have been consummated an eternity since, and physical stagnation would now be reigning. It is equally plain that the work which remains to be accomplished is a finite work, and is destined to be accomplished in finite duration.

What is the work with which terrestrial forces are occupied? What are the labors of oceans, and winds, and rains, and frost, and mountain torrents, and swollen streams, and pent-up fires? We witness here a grand antagonism of Nature's energies. While on one hand Nature has exerted herself to rear the continents, on the other hand a different set of forces has been equally assiduous in beating them down. There was a time when the igneous forces possessed the advantage, and island, and continent, and Alp rose triumphant over the sea. That was the age when the igneous forces were in their youth. Then all their elastic energies were commissioned to rear a dwelling-place for man. But, during geologic ages unnumbered, the powers of water have been wrestling with the powers of fire. Rains and floods have been tearing down what fire had built. The energies of fire have been wasting; the earthquake and the volcano have been stricken with the palsy of age. Old Ocean, however, is still in his youth. The volcano had been smitten with decrepitude even before the ocean had its birth. The denuding and destroying agencies of Nature have gained the ascendency, and, in the inevitable order of things, are destined to retain it.

Let us glance at the labors of water in leveling the inequalities which ancient volcanic energy had long ago created upon the surface of our planet. Throughout the whole extent of the circumambient sea, the tireless surge is gnawing at the rock-bound shore, and mouthful by mouthful the continents and the islands are being swallowed up. The sediment which every summer shower washes down the hill-side is so much material taken from the hill-top and deposited in the valley. The deep mould of the alluvial flat is made up of the spoils of the adjacent declivities. By as much as the valley is raised, the hills are lowered. The turbid waters of a winter stream are hurrying off with a freight of sediment stolen from a hundred townships. The mud which settles in my glass of river water upon a Mississippi steam-boat is a mouthful of the Rocky Mountains—or perchance of the Alleghanies—or, what is still more probable, it is a whole museum of soils, gathered from the fertile farms of New York and Pennsylvania—from the sandy cliffs of the Great Kenawha—from the clayey slopes of Cincinnati—from the slimy borders of Lake Pepin—from the melon-patch of a Cheyenne squaw, and from the beetling cliffs of the far-off Yellowstone. Of what part of the country is not this slime the washing? From month to month, and from year to year, and from age to age, this stately river is floating off the land—not noisily, but sullenly and angrily, as if the waters had some great wrong to avenge upon the land. And all these filchings from the mountain and the plain are restored again to the sea. Old Ocean is receiving back his own. The rivers are his allies, and right faithfully do they forage to supply the cravings of his insatiate maw.

We witness such work in progress during the brief moment of our tarry upon the earth. We look back along this line of operations, and discern for the first time the gigantic results which have already been achieved by the wearing agency of waters. Not during the lifetime of Adam's race alone, but during the age of quadrupeds which preceded him—through the dynasty of reptiles, still more ancient, have these denuding forces been ceaselessly engaged in scraping, and gouging, and scarring the face of Nature. River-beds have been deeply excavated and again obliterated by a plethora of rubbish poured forward by some more gigantic operation. Lake basins have been scooped out—Niagara gorges dug—square miles of land, with its underlying rocky floors, have been swept away. From the summits of the Catskill Mountains the Old Red Sandstone once stretched eastward perhaps to Massachusetts Bay. The powers of water have strewn it over Long Island Sound, and far to the seaward of Sandy Hook. The Cumberland Table-land once reached a hundred and fifty miles westward over the basin of Middle Tennessee. The site whereon the city of Nashville now stands was once a thousand feet beneath the level of the land. Half a state was scraped away to extend the borders of Mississippi and Alabama. The Alleghanies, in their prime, were three thousand feet higher than human eyes have ever seen them. Their ancient summits are sunken in the Atlantic and the Gulf of Mexico. The Great American Desert was once as fertile as the Valley of the Mississippi. A great river watered it for a thousand miles, while a hundred tributaries dispensed fertility throughout the region which was then the garden, as it is now the desert, of the continent. That fertile plateau has been drained to death. Each stream has drilled a frightful chasm deep through the rocky foundations of the plain (Fig. 96, 97). The mother stream, the Colorado, dwarfed to a withered mockery of what it was, now creeps along at the bottom of a narrow gorge whose rocky walls rise, in places, more than a mile in height. From the brink of this appalling chasm, three hundred miles in length, your vision struggles down six thousand feet into the realm of twilight; and in this prison the attenuated Colorado—patriarch of American rivers—is wasting its senile energies from year to year, but, with "the ruling passion strong in death," it is still carrying off the land, even though each season's work sinks it into a deeper grave.

Such are the works of running streams and corroding waves. The record of their labors is the utterance of the destiny of the land. History inverted becomes prophecy. The doom of the mountains is engraven upon their rocky buttresses. Half the pride of the Alleghanies has already been removed. Rounded hill-top is dissolving into plain. Defiant granite, which buffeted the lightnings that rent Sinai, and frowned upon the flood that drowned "the world," shall yet be brought down by the multitudinous pelting of rain, and the insidious sapping of frost. The mountains shall be wiped off. The continents shall be worn out. The rivers will have dug their graves. The ocean will have eaten up the land; and all there was of the dwelling-place of man will be a rocky islet, a ragged bluff, a sunken reef—the crumbs that fell from old Ocean's meal.

There was a time when, by degrees, the continents were

slowly and steadily surging from the sea. The sea, robbed of half his dominions, has ever since been raging around the borders of the land. At last he will again reclaim his own, and the universal empire will be Neptune's.

It is vain to hope that elevatory forces can permanently avert the disappearance of the land. We discover here another argument against the vague belief entertained by some, that the human fauna is to be succeeded by a higher one, as it has itself succeeded the lower. Should it be supposed that the ultimate submergence predicted is sufficiently remote to permit the interposition of a superior race of intelligences, I recall to mind the evidences that the lands are wasting and deteriorating; the river-beds are deepening, and diminishing the sources of irrigation; and all the populated regions of the earth are slowly approaching the desert condition of that ancient continent drained by the Colorado. Each continental surface in the geological succession is the exclusive gift to a single great fauna. A single race witnesses the disappearance of the freshness and fertility of the land. A new race would demand a thorough renovation, like that which immediately preceded the advent of man. Such a revolution the senescent forces are unable to inaugurate.

CHAPTER XXXVII.

THE REIGN OF UNIVERSAL WINTER.

WE open, now another volume of geological records. From this we glean another prophecy.

I have stated that the energies of the earth's internal fires are waning. There is a chain of effects which, when we trace them backward, conduct us to an ancient molten condition of the world. At a period comparatively recent, it was still so warm that tropical vegetation flourished within the arctic circle. At a remoter period, neither animal nor plant could endure the temperature which prevailed, nor the warfare which fire and water were waging with each other. We retain the solid monuments of a terrestrial condition which carries us still deeper into the heart of eternity; when the whole orb was a glowing ocean of incandescent lava, while yet the waters of the earth hung in invisible vapor upon the outskirts of the atmosphere, like a concealed foe meditating a secret attack upon a powerful enemy.

Few who have studied the physics of the globe, and fewer still who have deciphered geological records, doubt that such were once the temperature and conditions of our planet. From that state to this, it has passed by the simple process of cooling. We trace the footsteps of this progress at every stage. Through Azoic, Eozoic, Paleozoic, Mesozoic, and Cenozoic Ages, heat has been gradually wasted in space—the solid crust has been thickening—the surface conditions have been changing. The average temperature of to-day, instead of being a state that is destined to perpetuity, is but a passing phase; and when we shall have passed away with the other transient existences around us, some succeeding intelligence, gifted with the power to travel from sphere to sphere, will note the world in an altered condition.

I step here upon ground which has been somewhat contested. It was long since alleged that if our world be still in process of refrigeration, a sensible reduction in temperature ought to have taken place in 2000 years. But no such reduction has been satisfactorily established, though it will be confessed that we scarcely have exact observations on temperature which are more than two hundred years old. It was also alleged that since a reduction of temperature must be accompanied by a reduction of volume, the rate of the earth's rotation upon its axis must have been accelerated. But Laplace has demonstrated from ancient observations on eclipses that the mean day has not been diminished $\frac{1}{500}$th of a second since the time of Hipparchus, or during an interval of 2500 years. These negative results have been opposed to the theory of Cordier in reference to the former high temperature of the earth, and it has, till recently, been customary to speak of the thermal, no less than the astronomical conditions of our planet as constant. Poisson, an eminent French mathematician, proved, as was supposed, that the heat escaping from the earth in the latitude of Paris was only sufficient to elevate the temperature of a column of water eighteen inches high the trifling amount of one degree and a half. Vogt, a celebrated German geologist, affirms that the existing temperature of the surface of the earth is but one twelfth of a degree higher than it would be if the earth were completely cooled to the core. According to the later researches of Pouillet, the heat communicated to the surface of the earth from the central fire is but one fortieth the amount received from the sun; while, according to Fourier's celebrated computation, the heat radiated from the earth's surface is only sufficient to melt a layer of ice ten feet thick in one hundred years.

The most conservative of these results may be regarded as showing that our earth is actually losing heat to a perceptible and measurable extent. Neither is the amount of heat escaping at Paris to be taken as the measure of the reduction of the temperature of the mass of the earth in general. There are three hundred active volcanoes in existence, from the craters of which enormous quantities of heat are permitted to waste. The ocean, too, carries off vastly larger quantities than the land. The floor of the ocean is generally overlaid by a stratum of ice-cold water setting southward from the polar regions. This cold stream is overlaid by a warmer one moving northward from the tropics. Water being a better conductor of heat than atmospheric air, this cold stratum must necessarily abstract terrestrial heat with vastly greater rapidity than the average atmosphere of the temperate zone. Many observations indicate that the temperature of the solid crust beneath the waters of the ocean is much higher than that of continental surfaces, and hence imparts its warmth in larger quantities. Throughout all that part of the Frozen Ocean north of Europe and Asia, the temperature is found to increase at considerable depths, contrary to the well-known laws of hydrostatics. [See Appendix, Note X.] The same phenomenon has been observed on the coast of Australia, in the Adriatic, and Lago Maggiore. Horner asserts that in the deep soundings of the Gulf Stream, off the coast of the United States, the lead, when drawn up, "used to be hotter than boiling water."

These facts, with others, seem to demonstrate that our planet is wasting its warmth many times faster than the calculations of the mathematicians would indicate. It seems inevitable, therefore, that the earth should have expended sufficient heat in 2500 years to effect a sensible reduction in the length of the day.

Thanks to the mathematicians, they have again come to our aid. The tide-wave is a protuberance of the ocean-waters raised by the moon, and following the moon around the earth from east to west. This motion is *contrary* to the earth's diurnal rotation, and the friction of the tidal waters against the shore and the standing waters must necessarily tend to retard the rotary motion of the earth. Now it has been calculated that this retardation must have amounted to one sixteenth of a second in 2500 years. If, therefore, no counteracting tendency has been experienced, the sidereal day is one sixteenth of a second longer than it was in the time of Hipparchus. But Laplace has shown that the sidereal day has not varied in length. It follows, therefore, that the shrinkage of the earth from loss of heat has tended to accelerate its rotation to the extent of one sixteenth of a second in twenty-five centuries. Such an acceleration corresponds to a shortening of the diameter about sixty feet, and a reduction of the temperature of the whole mass of the earth one fourteenth of a degree.

When the earth was in its youth, just emerging from a molten state, the loss of heat and consequent contraction must necessarily have been rapid. During this period the sidereal day underwent a much more rapid shortening than at present. In the distant future, on the contrary, the loss of heat will become diminished to an extreme extent, and, as a consequence, the retardation caused by the tide-wave will gain the ascendency, and the day will eventually be lengthened to such an extent that the earth will always turn the same side toward the sun, as the moon always turns the same side toward the earth. The historic period of our race, as Mayer suggests, occupies consequently the comparatively brief space during which the retarding and accelerating tendencies neutralize each other.

These are the determinations of exact science. Mathematics have demonstrated that the cooling process which geology affirms of the past is certainly in progress in the present. It is immaterial how slow the process may be; the ultimate total refrigeration of the earth is a result which time will accomplish. Time, I say, since after the work is completed eternity will stretch onward as fresh, and inexhaustible, and limitless as when the career of planetary matter began.

This earth, to which our life-long round of labor and care is limited by an inexorable decree, was once a self-luminous orb. Far away in space, where Sirius was gleaming with his silver, or, perchance, his ruddy light, dwelt intelligent beings upon a planet which had already attained a habitable condition. From that abode the astronomer found means to contemplate the fiery globe that was destined to become the dwelling-place of man. Centuries of centuries later, the astronomer upon that distant orb noted the disappearance of a star upon which his predecessors had taken observations. Our planet had become opaque. Mists had gathered about it, and the ocean had descended from the clouds. Never more has this once resplendent orb greeted the eye of the astronomer of other systems; and while now the annals of his science perpetuate the memory of a lost star, that star first becomes a reality to conscious man. But *our* occupancy of the terrestrial globe is only a phase as evanescent as the self-luminous stage. While we build our cities and recount the achievements of a few generations past, this globe of matter hurries onward in its destined career as rapidly as a million years ago, when merely preparing for the occupancy of Adam's race. Every year and every day witnesses the dissipation of terrestrial warmth. While we ponder the great fact, the world is growing cold beneath our feet. The current of events is carrying us inevitably to a state of total refrigeration. Perhaps the mountains will have been leveled first, and the continents swallowed up in the sea. Perhaps the volcano will have been first extinguished, and the earthquake will have lain down to its final slumber. Buffon imagined that the final refrigeration of the earth would introduce the rigors of perpetual winter, and render our planet uninhabitable. Though more recent investigators have asserted that that event would only reduce our earth's surface temperature one fortieth of its present amount, it seems difficult to rest upon that conclusion. The interior of the earth is probably half as hot as the sun. The earth's molten core is separated from us by not more than a hundred miles of rocky crust. The glowing sun is a million times farther removed, and yet, it is alleged, yields forty times the warmth which we derive from the nearer heat. In face of the testimony of figures, it is scarcely possible to doubt that the final cooling of our earth will exert a greater influence upon its surface conditions than these philosophers have dreamed.

CHAPTER XXXVIII.

THE SUN COOLING OFF.

WE are not driven to the necessity of summoning exaggerated and imaginary agencies to the destruction of the earth. There are hostile powers reserved for the final conflict that will not be content with directing toward us merely "Quaker guns."

The sun, we say, affords us thirty-nine fortieths of all the warmth which we enjoy, and we feel quite unconcerned about the alleged slow cooling of the earth. To the sun we owe the numberless activities of the organic and inorganic worlds, and we feel quite independent of the waning temperature of this dying ember which we call the earth.

The amount of heat dispensed by our solar orb is truly something the contemplation of which overpowers the imagination. The rays which fall upon a common burning-glass, converged to a focus, speedily ignite a piece of wood. The heat which is received by a space of ten yards square is sufficient, as Ericsson states, to drive a nine-horse power engine. The amount of heat which falls upon half a Swedish square mile is sufficient to actuate 64,800 engines, each of 100 horse power. The total amount of heat received annually by the earth would melt a layer of ice one hundred feet thick. As the solar heat is radiated equally in all directions, it is easily calculated that the total emission of heat from the sun is 2300 millions of times the whole amount which reaches our earth.

Such an enormous expenditure of heat is sufficient to reduce the temperature of the sun two and one fifth degrees annually. During the human period of 6000 years, the temperature would have been reduced more than 19,000 degrees. At such a rate of cooling it is obvious that the sun must speedily cease to warm our planet sufficiently to sustain vegetable and animal life. But it is certain that the sun's high temperature has been maintained during almost countless ages anterior to the commencement of the human era. Those Titanic reptiles which could luxuriate only under tropical warmth flourished a hundred thousand years before the world was prepared for man; and those rank, umbrageous ferns, whose forms we trace upon the roof-shales of a coal mine, existed before the reptile horde, and purified the air for their respiration.

What unseen cause has perpetuated, for a million of years, those solar fires? Kepler asserted that the firmament is as full of comets as the sea is of fishes, and Newton conjectured that these comets are the fuel-carriers of the sun. Alas! we only know that the wandering comet, though flying in tantalizing proximity to the sun, but accelerates its speed and hurries onward, as virtue hastens past the vortex of ruin. Is it a chemical action which maintains the solar heat? The most efficient chemical action for this purpose is combustion. Now, if the sun were a solid mass of coal, its combustion would only suffice for the brief space of forty-six centuries to replenish the solar system with its vivifying influence. Is it the effect of the sun's rotation on his axis? Such rotation could generate no heat without the resistance of another body. Even if that other body were present, a calculation based upon the sun's mass and his rate of rotation shows that the heat generated could only supply the expenditure for the space of one hundred and eighty-three years.

There exists, nevertheless, a means of recuperation to the

solar energy. It is not an exhaustless resource, but it prolongs materially the period of the sun's activity. Though no comet has been *known* to fall into the sun, it is now generally admitted that cosmical matter is raining down upon the sun from every direction.

Besides the planetary and cometary bodies which revolve about the sun, it is now demonstrated that the interplanetary spaces are occupied by smaller masses of matter, from the size of a meteorite to particles of cosmical dust. These all are flowing about the sun in a circling stream, but forever approaching nearer and nearer, until they are gradually drawn into the solar fires. The showers of meteoric hail which pelt our earth at certain periods of the year are merely cosmical bodies that have been diverted from their path by the proximity of the earth in certain parts of her orbit. That faint cone of light which streams upward from the setting or the rising sun, near the time of the equinoxes, is but a zone of planetary dust illuminated by the sun's rays—a shower of matter descending upon the solar orb, and rendered visible to us, like the rain sent down from a summer cloud and projected upon the clear heavens beyond.

Arrested motion becomes heat. The blacksmith's hammer warms the cold iron. A meteorite falling through the earth's atmosphere develops so much friction as to generate heat sufficient to dissipate the body into vapor. One of these cosmical bodies falling upon the sun must, by the concussion, produce about 7000 times as much heat as would be generated by an equal mass of coal. It is thus that the enormously high temperature of our sun is maintained.

But the very mention of this source of recuperation of exhausted solar energy suggests a limit to the process. For how many ages can the cosmical matter within the limits of the solar system be rained down upon the sun without complete exhaustion? The space inclosed by the orbit of Neptune is not infinite. The supply of cosmical matter is but a finite quantity. Time enough will drain the bounds of the solar system of all its wandering particles of planetary dust. What then will be the fate of the sun?

The conviction can not be resisted that the processes going forward before our eyes aim directly at the final extinction of the solar fire. Helmholtz says: "The inexorable laws of mechanics show that the store of heat in the sun must be finally exhausted." What a conception overshadows and overpowers the mind! We are forced to contemplate the slow waning of that beneficent orb whose vivid light and cheering warmth animate and vivify the circuit of the solar system. For ages past unbounded gifts have been wasted through all the expanding fields of space—wasted, I say, since less than half a billionth of his rays have fallen upon our planet. The treasury of life and motion from age to age is running lower and lower. The great sun which, stricken with the pangs of dissolution, has bravely looked down with steady and undimmed eye upon our earth ever since organization first bloomed upon it, is nevertheless a dying existence. The pelting rain of cosmical matter descending upon his surface can only retard, for a limited time, the encroachments of the mortal rigors, as friction may perpetuate, for a few brief moments, the vital warmth of a dying man. The time is coming when the July sun will shine with a paler light than he now gives us at the winter solstice. The nations of men, if they still exist, will have emigrated from the temperate to the equatorial regions. New diseases will have diminished their numbers. Polar frost will have crept stealthily and steadily from Behring's Straits to the

Gulf of Mexico. Continental glaciers will again have brooded over the land. The prairie blossom will have perished beneath a mantle of snow as limitless as now the prairie expanse. The fluent rivers will have been chained to their rocky banks. The ruins of great cities will be bemoaned by wintry winds howling past in rage at the presence of unending frost. If yet a narrow belt remains where sickly verdure maintains the desperate conflict with the powers of cold, it is a dwarfed and arctic vegetation. The magnolia has given place to the birch. The cypress has been supplanted by the lichen-covered fir. The emerald has departed from the shivering leaf, and even the hardy violet is pale unto death. All things have assumed a faded and leaden hue. The Mongolian is not known from the Caucasian. Even the sooty negro, if he be not extinct, blanched from the want of light and heat, can only be recognized by his features. Pale, thin, and feeble, the shivering remnant of humanity have gathered themselves together into compact communities for economy of vital warmth. Forests are consumed to thaw the soil. Temples, costly structures—the patient rearing of the golden ages of the race—are pulled down to eke out the scanty supply of fuel. Men return to caves, whence they came in the beginning. Nature has become their enemy. Science and art are forgotten. The page which narrates the glory of the nineteenth century is like the narrative which tells us of the labors of the men upon the plains of Shinar. Year by year the populations become less—year by year the dread empire of frost is extended. Forests have been consumed; cities have been burned; navies have rotted in the deserted, ice-locked harbors; men have immured themselves in gloomy caverns till they have almost lost the forms of humanity.

The end arrives. Unless some sudden catastrophe shall

sweep the race from being in a day, the time will come when two men will alone survive of all the human race. Two men will look around upon the ruins of the workmanship of a mighty people. Two men will gaze upon the tombs of the human family. Two men will stand petrified at the sight of perhaps a hundred thousand corpses prostrated around them by the dire hardships which every moment threaten to carry them also away. These two men will gaze into each other's faces—wan, thin, hungry, shivering, despairing. Speech will have deserted them. Silent, gazing each into eternity—more dead than living—an overpowering emotion—an inspiring hope—and one of them drops by the feet of the sole survivor of God's intelligent race.

Who can say what a tide of reflections will rush for an instant through the soul of the last man? Who shall listen to his voice, if he speaks? On whose ear shall fall the accents of his sorrow, his wonder, or his hope? Thrice honored, thrice exalted man! He stands there to testify for all mankind. On him has been devolved the unique duty of uttering the farewell of our race to its ancient and much-loved home. In what words will he say farewell?

The last man has composed his body to eternal rest. The once fair earth is a cold and desolate corse. Nature's tears are ice; she weeps no more. The face of the sun is veiled. It is midnight in the highways of the planets. The spirits of heaven mourn at the funeral of Nature.

Let not the reader be distressed at this picture. The last two men will be neither our children nor our children's children. Our thoughts have been wandering through cycles of years. The clock of eternity ticks not seconds, but centuries. We shall not anxiously measure the sun's intensity from day to day, nor from year to year, lest we be able to discover his waning strength. The embers of a bonfire will furnish warmth for the lifetime of an ephemeron. A molten lava-stream consumes a hundred years in cooling. The great globe of the earth, which is cooling now at the rate of a degree in thirty-five thousand years, was once a sphere of molten granite, and has consumed time enough to pass from that state to this. The sun is so vast that, though he began to cool at a still remoter epoch, the temperature retained to-day is 46,000 times as high as that of the surface of our planet. The epoch when his rays will be sensibly weakened is at a distance expressed by millions of years.

What thoughts rise upon us as we utter these words! We hang here upon our planet, poised in the midst of infinite space and infinite time. Whence we came, we know not; whither we are bound, hope and faith only can reveal. We open our eyes for a moment, like an infant in its sleep, and anon they are closed; or, perchance, like the waking somnambulist, in his fall from the house-top to the pavement, we rouse to an instant's consciousness of the rush of events and the coming crash—and the busy activities of Nature move on as if we had not existed.

A few days since, a friend of mine exhibited to me a silver coin dug up from the rubbish of the hoary East. It was rude, irregular, and begrimed with age. Upon one side was raised the image of a Grecian warrior. Above the head I could trace, with difficulty, but with certainty, the Greek letters which spelled the name of Alexander. Venerable coin! thought I; and my imagination wandered back through twenty-three centuries, till I saw the Issus and the Granicus, and the hosts of Darius melting before the fury of the Macedonian conqueror. I felt transported back to antiquity. But then I remembered the Nineveh marbles upon which I had gazed, and the black and skinny mummies that had looked out at me from their withered eyeballs, and imagination spanned another interval of ages; and I stood upon the banks of the Tigris and the Nile, and the forms of Sennacherib, and Menes, and Mŏses passed before me. As chance would have it, I returned, and, passing through a cabinet where the "medals of creation" had been ranged in regular order, the ponderous molars of an extinct mammoth, dug from the soil of Michigan, awakened a new thought. By its side rested the skull of Oreodon, with its sheep-like teeth in a hog-like head; and, being in a mood for revery, I thought of the distant Missouri plains where Oreodon had grazed; and of the vast lake—thrice the size of Superior—from whose water he had drank, and on whose muddy banks had crawled turtles that could carry oxen on their backs. And then I remembered that thought had darted back over another stretch of ages to a time when God had not yet said, "Let us make man," when the wide continent was the pasture-ground of elephants, and mastodons, and wild horses, and camels, and sloths, and quadrupeds of strange shapes which were blotted out of existence before ever human eye had gazed upon them.

Here, I thought, are the relics of a genuine antiquity. I sauntered on, and the teeth, and vertebræ, and dimly-outlined forms of Ichthyosaurs, and Deinosaurs, and flying lizards, and fishes clad in mail—bucklered and helmeted fishes —these in succession passed before my eyes. And then winged thoughts flew back through those dim ages of the world's history which we call Mesozoic. I breathed a stifling atmosphere; tepid vapors rose all around me; strange foliage fringed bayous of which I had never heard; neither bird nor insect stirred the fervid atmosphere; there were no forests; the continents were but just arising from their sea-couches, and no footprint had yet been impressed upon their slime-covered heads. And then I thought again of the silver coin which bore the image and superscription of Alexander, and wondered why I had called it venerable. Why? since twenty populations had possessed the earth, since the relics of those bucklered fishes had been animate, and this coin—why, it had been stamped in the last part of the lifetime of the twentieth population; and there were nineteen before it which had become extinct.

And so my feet were lifted up from earth; I was pillowed upon a bright cloud, and floated in eternity. And I saw the long history of the world I had left stretching backward from the spot where I had left it, till it vanished from view, like the track of a railroad on the boundless prairie. With the flash of a thought, I pursued it over millions of ages, till I saw it dissolved in fire—till luminous vapors rolled up and rested upon the bosom of infinite space. In this cloud of fire the track of terrestrial history lost itself, and I dared not plunge through the flame in search of a beginning.

Then I thought, here at length is the dwelling-place of antiquity. What is this which men call ancient and venerable? Would that the scales could be removed from our eyes! Would that the fog would lift, and men could once look out upon the magnitude of the universe—the majestic span even of terrestrial history—the might, the greatness, the wisdom, the glory of that Intelligence which, at a glance, takes in all space, all time past, and all time to come!

CHAPTER XXXIX.

THE MACHINERY OF THE HEAVENS RUNNING DOWN.

LET the earth have frozen; let the bright sun have been extinguished; let the moon and stars "wander darkling in the eternal space." Will this, then, be the end of matter's history? Is this the consummation of which philosophers, and poets, and patriarchs have dreamed and prophesied? From the pinnacle on which we stand we can discern the course of Nature still wending onward. There must be progress even after the funeral of the sun. As that bright luminary shines on after the fall of generations of men—as he shines serenely and undisturbed even in dead men's faces, so will gravitation continue to prosecute its work even among the corpses of planets and suns.

Hark! from the highways of the comets come tidings of friction in the machinery of the heavens. The filmy wanderer encounters resistance in his long journey to the confines of the solar system. He plows his way through a resisting medium. The balance of centripetal and centrifugal forces is destroyed; the central attraction preponderates; he falls toward the sun; his orbit is diminished; his motion is accelerated, and he comes back to his starting-point earlier than the time which astronomy had appointed. Here we get the first disclosure of the existence of a subtile material fluid pervading space.

This remarkable retardation was first observed in the successive returns of Encke's comet. This comet has at present a period of about 1210 days, and it returns each time two hours and forty-five minutes sooner than calculation requires. Since 1789, its period has shortened two days and sixteen hours. Similar retardation—resulting in a similarly accelerated *angular* velocity—has been fully established in the cases of the comets of Brorsen, Faye, and D'Arrest.

The only explanation that has ever been offered of these exceptional phenomena is the assumption of the existence of an all-pervading resisting medium commonly called ether. Since the undulatory theory of light became established, the existence of such a medium has been recognized as necessary, and its presence has been assumed throughout all those realms of space to which light has penetrated.

A belief in the existence of an all-pervading ether is much more ancient than the observations which have made it a scientific datum. In the astronomy of the Brahmins, the stars are said to swim in ether, as fishes in the water. Kepler supposed comets to be native inhabitants of this ethereal medium, like fishes in the sea. The Cartesian doctrine of "vortices" presupposes an all-pervading material fluid. The existence of such a fluid was admitted by Newton; and he demonstrates that its tenuity must be greater than that of our atmosphere at the distance of two hundred miles from the earth. In more recent times this doctrine has been maintained or admitted by Whewell, Sir John Herschel, Thompson, Mayer, Littrow, Helmholtz, Grove, Tyndal, Watson, McCosh, Comte, Rorison, and, in short, by every physicist who has investigated the subject.

We are compelled, then, to assume the position that a resisting fluid permeates space, and that the heavenly bodies do *not* move in a vacuum. The consequences of this admission are stupendous beyond conception. Laplace demonstrated that if the planetary bodies are solid, and if they move *in vacuo*, their mutual perturbations, in long cycles, compensate each other, and the stability of the solar system is perfect. The contingent part of this proposition possesses all the significance. Neither are the planetary bodies solid, nor do they move *in vacuo*. The effect of the terrestrial liquids is apparent in a considerable lengthening of the sidereal day—which, for the time being, is counterpoised by the shrinkage of the earth—while the effect of the resisting medium has been wrought out in the partial arrest of the whole brood of comets.

The retardation of Encke's comet is such that it would lose one half of its present velocity in 23,000 years. A power which can sensibly check the flight of the filmy comet can also retard, however minutely, the motion of the ponderous planet. Jupiter, by far the largest of the planets, would lose one thousandth of his velocity in seventy millions of years. The length of the period has nothing whatever to do with the result. If the motion be inevitably and perpetually toward precipitation into the sun, the event is as demonstrable as the fall of an aerolite to the earth. Not only are the cloud-like comets slowly approaching the sun in spiral curves, but every revolving planet—every material particle in the solar system—is borne forward by the same unalterable decree. It is the presence of a resisting ether which conditions the precipitation of that meteoric rain which retards the cooling of the sun. The fall of comets and planets to the sun will still farther delay the final refrigeration of that luminary, without averting it.

The proof of the existence of a resisting ether in space has disclosed the decree which records the doom of the solar system. Whewell says: "Since there is such a retarding force perpetually acting, however slight it be, it must in the end *destroy all the celestial motions.* * * * The moment such a fluid is ascertained to exist, the eternity of the movements of the planets becomes as *impossible as a* *perpetual motion on the earth.*" Helmholtz says: "A time will come when the comet will strike the sun; and a similar end threatens all the planets, although after a time the length of which baffles our imagination to conceive it." Mayer contemplates the precipitation of asteroidal and planetary masses upon the sun. Comte says: "In a future too remote to be assigned, all the bodies of our system must be united to the solar mass, from which it is probable that they proceeded." Rorison, defending the Mosaic account of creation, admits, speaking of the earth: "It was once all nebula; it will yet, if left to physical agencies, collapse into an exhausted and extinguished sun." Watson says: "If we grant that the retardation of the comets arises from the existence of an ethereal fluid, the total obliteration of the solar system is to be the final result."

This, then, is the conclusion of science. So far as we have been able to acquaint ourselves with the laws which regulate the movements of the planetary bodies, the duration of the present order of the solar system *is finite.* Nothing but an *infinite miracle* can save it from destruction. That such a miracle will be wrought we have no warrant for assuming. From the very beginning of its career, so far as we can judge, the history of matter has been wrought out in accordance with methods which we style the "laws of Nature." These methods have never been abandoned, and there is not a particle of evidence furnished by science that they ever will be abandoned until they shall have completed their work. Whether the forces of matter be viewed as inherent powers or as "immediate divine agency," the argument from induction, in which the doctrine of a final catastrophe rests, is an argument possessing strength beyond the power of arithmetic to express.

It is true that the final catastrophe is removed to the

distance of millions of millions of years. It is true that the sun may have cooled millions of years before the consummation of the final crash. It is true that a million of years before the cooling of the sun the earth may have become desolate and tenantless, as it was a million of years before it received its first inhabitant, or as the moon is to-day, poised in space before our very eyes. It is true that thousands of years may yet elapse before the ordinary powers of geology shall have leveled the continents, or changed their habitable conditions to such an extent that man and other organic beings will have passed away. But the magnitude of the numbers by which these intervals of time are symbolized does not embarrass the argument. Infinity dwells not alone in years. Are the recitals of Astronomy less fanciful than these? Are the data with which she deals less staggering to the human mind? We think ourselves dwelling in the immediate neighborhood of the sun, since, perchance, his light comes to us in eight and a half minutes. Yet his distance is such that a traveler, setting out for the sun by railway on the day of his birth, and traveling continually thirty miles an hour, would attain the age of fourscore before having spanned one fourth of the vast interval. Were he and his posterity to complete the journey, and were generations to succeed each other according to the established rule, the twelfth generation would appear before the station should be reached. The great luminary would be pressed by the foot of his great-grandson's great-grandson's great-grandson, and he would be upon the tottering verge of fourscore. Had Christopher Columbus set sail for the sun instead of a new continent, and traveled continuously at three times the speed of a steam-ship, he would only have reached his destination this year. This is the distance which light travels over in eight and a half minutes. There is no doubt admissible in regard to the distance of the sun from the earth, or the velocity of light, and yet the nearest of the fixed stars is so remote that its light has consumed ten years in passing to our earth; and there are visible stars so distant that their light has occupied the lifetime of our race in darting over the measureless void. In each second of that interval it has traveled a distance measured by seven times the circumference of the earth. Nay, I may gaze through the telescope on any star-lit night, and gather into my eye rays which set out from a distant nebula ages before even the race was called into being whose slowly-developing science has enabled me to make these calculations and gather up this feeble light.

These are values which the positive science of astronomy affords us. Nor are the wonders of physics less overwhelming. The amount of heat sent off from the sun in one minute is, according to Mayer, 12,650 millions of "cubic miles of heat." Now what is a cubic mile of heat? In the conventional language of the physicists, it is the quantity of heat necessary to raise the temperature of a cubic mile of water one degree Centigrade. Have we any conception of the amount of heat required to do this work? In order to subdivide the quantity till we reach a limit which our intellects can grasp, let me state that one cubic mile of heat contains 408 billions of units of heat; and a unit of heat is the quantity of heat required to raise one kilogramme—or about one quart—of water one Centigrade degree, which is one and four fifths degree of our scale. In other words, then, the sun emits more than five septillions, or five thousand millions of millions of units of heat every minute. In a year the amount is 522,000 times as great; and in the brief duration of our race it has been more than three thousand million times seven septillions of units of heat. These, let the reader remember, are the data of exact science. They are not the millions of years in which the geologist symbolizes the age of the world.

Nature thus, on every side, launches us forth upon the borders of infinity. We flutter about like insects on a flower-bed, and stand awed before the "boundless prairie," the "primeval forest," or the "shoreless ocean." We speak of planetary distances and stellar pathways, but our efforts to compass thoughts like these are as the navigation of the paper nautilus upon the heaving bosom of the broad Pacific.

And yet such quantities are not imaginary. Such intervals as millions of ages *will be passed;* such intervals *have already passed.* To the eye of that all-comprehending Intelligence whose works these are, whose plans these are, millions are but molecules in the constitution of a universe; the lifetime of a planet vanishes as a thought. To a being who is Infinity, the very units of measurement are infinity; one stroke of the hand is infinite space, one step of progress is infinite time.

Where, then, is perpetuity? The untutored savage looks upon the ancient forest as all-enduring. His fathers sat beneath the shade of the self-same tree as stretches its arms above his own squalid hut. The poet sings of the "eternal hills," or fancies that in the ocean he discerns "the image of eternity." The philosopher thought he had demonstrated that at least the solid earth should endure forever, and the coterie of planets should not cease to waltz about their sun. But at length we discover not only that forests appear and disappear—not only that the mountains crumble away from age to age, and Old Ocean himself has limits set to his duration—but even yonder burning sun is slowly waning, and the very earth is wearily plodding through the mire of ether, and we can foresee the time when, with all her energies wasted, the fire of her youth extinguished, her blood curdled in her veins, her sister planets in their graves or hurrying toward them, she herself shall plunge again into the bosom of her parent sun, whence, unnumbered ages since, she whirled forth with all the gayety of a youthful bride.

Such is the position to which science conducts us. We feel that we stand here upon sure foundations. We have no means of measuring accurately the length of eternity's years, but we know they exceed ours a million-fold. We can clearly translate the watchword of the hosts of space. "*Not for perpetuity*" is written upon every lineament of the solar system. We contemplate the matter of the system aggregated into a cold and blackened mass at the centre. No more sun, no more planet, no more satellite, no more comet, or metorite, or zodiacal luminosity, but winter, and the silence of death, and the darkness of Nature's midnight, penetrated only by starlight, whose maternal source may even then have been blotted out—a solitary grave upon a distant plain, in the midst of the howling desolation of an arctic winter.

But imagination, indefatigable, with wing unwearied while yet there remains another height to scale, pausing here but an instant, throws her glances still beyond. Into that remoter eternity which stretches still beyond the sepulchre of the solar system her vision penetrates. Shall we venture to delineate the vicissitudes which she sees transpiring in that deeper depth? They are the figures of things but faintly limned against the curtains of infinity. But yet there is no religion which forbids us to reproduce that ethereal vision. Let us exhaust the revelation of Nature, and seize upon knowledge which lies next door to the supernatural world.

Astronomy calls every star a sun, and declares that our solar orb is but one in a firmament of suns. When we gaze at night upon the stellar host we descry the nearest members of a cluster of suns, which, vast as it is, has limits which have been surveyed. Sir William Herschel, with the graduated powers of his great telescope, sounded the depths of the firmament, and determined its extent in every direction. In the midst of this circumscribed cluster of suns our solar luminary holds a position.

Beyond the confines of the outermost zone of stars lies an empty void. Sir William Herschel, with the higher powers of his instrument, looked through the loop-holes of our firmament, and sent his vision across the cold and desert space which spreads out on every side. The cheering starlight that had accompanied every farther stretch across the populated fields of our firmament now forsook him, and he gazed only upon dread emptiness and blackness. For a moment he imagined he had caught a glimpse of infinity; but lo! across that measureless void appears another firmament! And still other firmaments, on every side, beam on us with a blended gleam which fuses their constituent suns into a cloud. These are the nebulæ.

To what order of distances are they removed? Are their histories identical with the history of our firmament? Is infinite space occupied by an endless succession of such starry clusters? These are questions which we shall find answered when thought is permitted to penetrate one step farther, and set foot within the bounds of the supernatural world.

These systems of suns, with their probably attendant planets and satellites, all exist *under one constitution.* The spectroscope has demonstrated that the light of the different heavenly bodies is substantially identical. It has demonstrated the identity of the *luminous matter* of sun, and comets, and stars, and nebulæ. It has declared the existence of carbon in the comets, of hydrogen, potassium, magnesium, and sundry other bodies in the sun, and of several of the same in the fixed stars and the nebulæ. There is little uncertainty about these determinations. Like an expert who identifies the handwriting of a criminal when he who penned it may have fled a thousand miles away, this little instrument, by an analysis of light that has wandered a thousand years away from its source, declares the nature of the luminous body that sent it forth. *One sort of matter* exists throughout all the wide realm which human vision has traversed.

One *ether* extends through all. The conditions under which light is propagated are identical upon Sirius and in the apartment lighted by a jet of gas.

Gravitation reigns over all. The phenomena discerned in the motions and phases of the stars and the nebulæ are such as attest the dominance there of the same law as holds the earth in its orbit, or guides an apple to the ground.

The law of *rotation* reigns over all. From the revolving moon to the vicissitudes of the variable stars—winking night and day in regular alternation—and even to the spiral nebulæ whose stupendous gyrations have given shape to their flowing vestments—every where the equilibrium of celestial bodies, like that of a top or a gyroscope, is maintained by rotation.

Our own firmament of stars, which we are not permitted to view as we view the nebulæ, from a distance, reveals even to a beholder from within the fact of its rotation, as the progress of a sloop upon a river is revealed by the apparent motions of the trees upon the banks. Mädler has announced the discovery of the astounding fact that the entire firmament is describing a slow and majestic gyration about a centre which, to us, seems located in the Pleiades. In this common whirl of a million suns our sun participates, and, with his retinue of planets, moves forward through space at the rate of one hundred and fifty millions of miles a year. And yet so vast is the circuit upon which he is launched that 18,200,000 years will have faded away before he shall have completed a single revolution.

If throughout all these boundless intervals of space a resisting ether is present—if it be a fact that a material fluid pervades all the wide realms which light has traversed, what is the conclusion which looms up as a consequence? Are we not compelled to recognize the fact that every sun in our firmament, as it journeys round and round in its circuit of millions of years, is slowly, but surely as Encke's comet, approaching the centre of its orbit? And in that most distant future, the contemplation of which almost paralyzes our powers of thought, is it not certain that all these suns must be piled together in a cold and lifeless mass?

I forbear to say more. With reverence I refrain from the attempt to lift the veil which conceals the destiny of other firmaments. I dare not hazard the inquiry whether an immensity of firmaments may not be executing their grand gyrations on a still larger scale; and whether these, in turn, may not be destined to a grander cosmical conglomeration. It is vain to push our conjectures farther. We have even here entered upon the border-land of infinite space. In the presence of Infinity, what can man do but bow his head and worship?

Reason assures us that somewhere the tendency to central aggregation must be stayed. A universe of worlds can never be gathered together in a single mass. Within the bounds of the visible we see all matter wending its way toward centres of gravity. Within the bounds of our firmament we see all matter tending toward one centre. Let us content ourselves to speak of this. This shall be our universe. This is the universe whose final aggregation into one mass we are compelled to contemplate.

CHAPTER XL.

THE CYCLES OF MATTER.

OF what, now, is this stupendous result the consequence? This is the goal toward which, for millions of ages, the forces of matter have been struggling. During every moment of this long history, gravitation has striven to draw these myriads of worlds together. They have embraced each other at last, and gravitation has retired to slumber. While yet these worlds were in active life, every sun was a heated globe, dispensing warmth through infinite space; and every planet may have been the seat of life, enjoying the boundless munificence of heat. As long as heat remained to be dispensed to planetary orbs, they were the seat of all those myriad activities of which solar heat is the origin and source—currents in the atmosphere and in the waters—ascending vapors and descending rains—the nurture of vegetable and animal life and motions—the disintegration of continents, and the strewing of ocean bottoms with layers of sediments for the upbuilding of new continents. Through numberless interactions of heat, and electricity, and light, and magnetism, and mechanical forces, and chemical affinity, the web of material and organic history was woven. The equilibrium of the heat of our universe has now been attained. No farther interactions and transformations can ensue. Every particle of matter is equally cold. Every corner of space is equally dark. The electricities that have been worried with disturbances and divorces innumerable are now firmly locked in each other's embraces. Every chemical element has united with its first choice. There can be no farther decompositions or recompositions. The forces of matter have spent themselves. After a fierce conflict, they lie mutually slain, upon a long-contested battle-field. The struggle is ended—nothing stirs—night comes down and casts her pall over the corse of matter.

From this exit of material existence we shrink back to the times in which we live, and inquire, What are all the myriad activities of the passing world—what are rolling tides, and surging waves, and ocean streams—what are mountain births, and volcanic eructations, and continental throes—what are wasted lands, and Niagara gorges, and ocean sediments—what are worn-out continents, and extinguished populations, and terrestrial revolutions—what are all these vicissitudes through which the earth has passed, and all these phenomena which to-day are transpiring—what are they all but the *incidents* attending the progress of the active forces of Nature toward their destined equilibrium? In their restless and active lifetime they show themselves under myriads of guises, and work out their myriads of incidents; but the great law which is over them hurries them ever onward in but one direction, and the end of that is equilibrium, stagnation, death.

Is this, then, the end of matter? Is it for this that space has been populated with worlds innumerable? Was it for this brief ferment that a past eternity should brood over nothingness, and an eternity to come should ache with the recollection of creation foiled? The forces of matter can do no more. The machinery of the universe has run down. Beyond and above is only the Eternal Omnipotence. There is now no power in the universe but Deity. When he wills the resurrection of matter shall dawn. New life will thrill through every vein of the ancient corse. When he wills the forces of matter shall hie again from their hiding-places.

Heat will again be gathered into central masses. Matter will dissolve into liquids—liquids burst into vapor, and fill again the vault of space—cohesive affinities will be sundered—chemical unions will be unlocked—electrical and gravitating forces will resume their play, and once more will begin the long series of activities which make up the lifetime of firmaments, and systems, and worlds. The matter of our solar system—or of a system like ours—will again be isolated; the endless whirl of fiery vapor will detach rings, in succession, which will consolidate into planets and satellites—another earth will spring up—another period of the reign of fire will ensue—and then another reign of water—and then another long line of organic creations will begin, and, in due time, in some distant future age, another intelligent race will populate another earth, and dream, as we now dream, of the beginning *whence*, and the goal *whither* the grand rush of events is carrying them. This is one of the Cycles of Matter.

In what light, then, are we to regard all the vicissitudes and activities of the lifetime of a universe? What are they but a brief agitation on the surface of the infinite ocean of matter—a momentary ripple raised by the presence of the Omnipotent hand—destined speedily to subside, and again to be raised by the breath of Omnific Power?

In the presence of such conceptions as these, what is man, and what are the works of his hands? What are fleets, and forts, and cities with their insect hum? What are temples, and pyramids, and Chinese walls? The agitation of particles of dust in a distant corner of the universe. The track of an insect on the ocean's shore. The breath of an infant in the tornado's blast.

But what is the spirit of man, whose thoughts thus wander through eternity? What is the intelligence of man which climbs the battlements of the palace of Omnipotence

—which seizes hold on infinity—which, though chained in flesh, spurns its fetters, and feels evermore that it is the offspring of God—the brother of angels—the heir of perpetuity—and will soon shake its shambles down amongst the rubbish of decaying worlds, and dwell superior to the mutations of matter and the revolutions of the ages? What, in comparison with the crumbling of mountains and the decay of worlds, is the being possessed of such a consciousness and such a destiny? Who shall tremble at the wreck of matter, when, in perpetual youth, he shall outlive suns, and systems, and firmaments, and through the ceaseless cycles of material history shall see creation rise upon creation—the ever-recurring mornings of eternal life?

APPENDIX.

NOTE I., page 50.

The doctrine of the central igneous fluidity of the earth is generally accepted by geologists. The hypothesis of intense chemical action as the cause of existing internal heat—a hypothesis first enunciated by Sir Humphry Davy*—is, however, revived from time to time under some novel modification. Dr. T. S. Hunt, while admitting the primordial incandescence of our planet, has maintained, in a series of lectures before the Lowell Institute of Boston, that the solid crust is probably not less than 2000 miles in thickness, and envelops a solid nucleus, with a comparatively thin belt of material between the two, which has been reduced to a soft and pasty condition by the combined action of heat, water, and chemical affinity. In his lecture on "Primeval Chemistry," more recently delivered before the American Institute in New York, he is reported as saying that "the earth must have a crust several hundred miles in thickness;" that "granite is in all cases a secondary rock, derived from sediments crystallized through the agency of water and heat;" and that "the theory which ascribes volcanic products to the supposed uncooled liquid centre fails entirely to account for the great diversity in composition of these products, all of which, wherever found, are represented in rocks of aqueous origin."

Mr. N. S. Shaler has attempted to show, in an ingenious paper read before the Boston Society of Natural History (Proceedings, vol. xi., p. 8), that the solidification of the earth began at the centre and proceeded toward the periphery—that finally solidification began at the periphery and proceeded toward the centre, leaving, within the era of recognizable geological events, but an insignificant portion of the earth in its primordial fluid state.

* Unless, indeed, Milton can be said to have first suggested it in the following words:

"The force
Of subterranean wind transports a hill
Torn from Pelorus, or the shattered side
Of thundering Ætna, whose combustible
And fueled entrails thence conceiving fire,
Sublimed with mineral fury, aid the winds,
And leave a singed bottom all involved
With stench and smoke."—*Paradise Lost*, i., 230.

Professor James Hall, also, in his recent lecture before the American Institute on the "Evolution of the American Continent," is reported (in the New York Tribune) as advancing views which indicate that he has relapsed into the ranks of the most radical Neptunists. "I desire," says he, "to impress upon you this one truth, that we have not, in our geological investigation, succeeded in going back one step beyond the existence of water and stratification—one step toward this so-called primary nucleus of molten matter. So far as we have any knowledge of the materials in the interior of the globe, they appear to us only as trap dikes, and these occupying only a very small area upon the surface. This original nucleus that has been talked about in geology has produced no effect upon the surface of the earth; neither upon its mountain chains or any other of the great features of the continent."

"This idea of a great primary nucleus is only theoretical. It has not in it any thing tangible. The earliest rocks of which we have any knowledge were deposited by the ocean, under conditions similar to those which now exist. The conditions of the ocean currents are the same now as they have been from the earliest time. From the earliest history of the American continent—from the earliest history of any other, we know that the ocean currents have prevailed as they now prevail, moving northward and southward; and here, at least, the transporting power has generally been from the north toward the south and west; and we have abundant evidence that all the materials composing our continent have been derived in that way from the transporting agency of currents of water alone."

Professor Hall seems to have taken the laurels from the brow of M. Comte in his resignation to the consequences of the Positive Philosophy. There are many positions in the foregoing quotation which are destined to be shaken as by an earthquake shock generated by those very internal fires which he so irreverently ignores. Though this is not the place for argument, I will not refrain from reminding the reader that if our world has been cooling for many ages, as science demonstrates that it is cooling to-day, there must have been a time when the *first* aqueous sediments accumulated. What was *their* origin? I very well understand that the reply will be, that we neither *know* that the earth has been in process of cooling from a high antiquity, nor have we *seen*, except in isolated patches, the supposed foundation-lavas and granites which constituted the primordial crust. When I stand by the Michigan Central Railway, and see the "Blue Line" freight-cars pass, bearing the inscription "Great Central Route; through freight from New York to the Mississippi," I should consider it folly to deny that these cars have proceeded from New York, and base my denial on the fact that I had *never seen them at that point.* It is thus that *events* rush past us, and he who will read the legends which they bear may learn somewhat both of the beginning and the end. Lack of demonstration is not necessarily nescience. It is too much the fashion of a certain school to apply the shears of nescience to scientific and philosophic systems, to crop and prune them to predetermined shapes. Whither the known points us, let us follow; and if we can not discern things clearly, let us be content to see them "through a glass darkly." It would be stupidity to ignore the existence of a solar orb even in total eclipse.

This revulsion in the popular view has to some extent been produced by the weight of well-known names recorded against the doctrine of primordial fusion and continued central fluidity. Sir David Brewster denounces "the nebulous theory" as "utter nonsense;" and Mr. Evan Hopkins has publicly denied the accepted doctrine of a slow increase of temperature in penetrating toward the earth's centre. It is certain, however, that the facts upon which his denial rests have been generated by abnormal and perturbating influences. Mr. W. Hopkins several years since contended that the solidification of the earth must have begun at the centre, simultaneously with the formation of the superficial crust. Sir Wm. Thompson maintains that the rigidity of the earth is required by the phenomena of precession and nutation. Against these conclusions, however, Delauny very recently opposes the results of experiments which show that a body of water inclosed in a rotating glass globe promptly partakes of the rotation of the globe, and becomes physically a part of it. The author remains decidedly of the opinion that the balance of evidence sustains the doctrine of central fluidity. The reader who desires to examine farther the objections urged against this doctrine may consult Hopkins (Wm.), in Phil. Trans. of the Royal Society, 1836, p. 382; also 1839-40-42; also Quar. Jour. Geolog. Soc., Lond., vol. viii., p. 56; Thompson (W.), on the Rigidity of the Earth, in Proceedings Roy. Soc., vol. xii., p. 103; Tyndall, in Fortnightly Review. On the subject of mountain-formation, see Hall (James), Paleontology of New York, vol. iv., Introduction; Dana (J. D.), Address before the Amer. Assoc. for the Advancement of Science, Providence, 1857.

Note II., page 71.

As the recent discovery of traces of animal life two whole systems lower in the series of strata than had heretofore been known is an event of extraordinary importance in the progress of our knowledge of the world's preadamic history—insomuch that Sir Charles Lyell characterizes it as the greatest geological discovery of his time—I introduce here a somewhat complete series of references to the papers which have been published on the subject:

1858, May. Hunt (Dr. T. S.), Remarks on the presence of iron ores and graphite in Laurentian strata as affording evidence of the "existence of organic life even during the Laurentian or Azoic period." Amer. Jour. Sci. and Arts [2], xxv., 436.

1858, Oct. Logan (Sir W. E.) received the first specimens of suspected fossilferous rock from the Grand Calumet on the River Ottawa, collected by Mr. J. McMullen. Figured in "Geol. of Canada," p. 49, and in numerous places since.

1859, Jan. Hunt (Dr. T. S.), Reiterates his convictions as to the "existence of an abundant vegetation during the Laurentian period." Quar. Jour. Geol. Soc., Lond. Reviewed Amer. Jour. [2], xxxi., 134.

1859, Aug. Logan (Sir W. E.), Exhibits specimens before Amer. Assoc. Adv. Sci. at Springfield.

1859. Logan (Sir W. E.), Publishes notice of the above. Canadian Naturalist, iv., 300.

1861, May. Hunt (Dr. T. S.), Sets forth more fully his views on probable Laurentian life. Amer. Jour. Sci. [2], xxxi., 396.

1862. Logan (Sir W. E.), Exhibits specimens in Great Britain which are held by Ramsay to be organic.

1862, Dec. Dana (Prof. J. D.), Suggests indications of organic life in Azoic rocks. Manual of Geology, p. 145.

1863. Logan (Sir W. E.), Describes and figures specimens from Grand Calumet. Geology of Canada, p. 49.

1864, Mar. Logan (Sir W. E.), Preliminary notice of additional specimens discovered by James Lowe in Grenville. Amer. Jour. [2], xxxvii., 272.

1864, May. Hunt (Dr. T. S.), Preliminary notice of some specimens. Amer. Jour., xxxvii., 431.

1864, Nov. Sanford (Mr.), Announces Eozoön in Connemara marble of the Binabola Mountains, Ireland. Geological Magazine. Reannounced in "Reader," Feb. 25, 1865.

1865, Feb. The Quar. Jour. Geol. Soc., Lond., contains the following papers:

Logan (Sir W. E.), "On the occurrence of Organic Remains in the Laurentian Rocks of Canada."

Dawson (Dr. J. W.), "On certain Organic Remains in the Laurentian Limestones of Canada." A microscopical investigation and determination of zoological relations.

Carpenter (Dr. W. B.), "Notes on the Structure and Affinities of *Eozoön Canadense.*" Farther microscopical descriptions.

Hunt (Dr. T. S.), "On the Mineralogy of *Eozoön Canadense.*"

The foregoing papers were republished in the Canadian Naturalist [2], ii., 92, April, 1865; also in a pamphlet entitled "On the History of Eozoön Canadense," April, 1865; also, with remarks by the editors, in Am. Jour. Sci. and Arts [2], xl., 344, Nov., 1865.

1865, April. Jones (Prof. T. R.), Discusses the geological and zoological relations of Eozoön. Popular Science Review.

1865, May. Carpenter (Dr. W. B.), Further discusses Eozoön. Intellectual Observer (two plates).

1865, June 10. King and Rowney (Profs.), Question the organic nature of Eozoön, while Dr. Carpenter sustains it, supported by the authority of Milne-Edwards.

1866, Feb. 10. Carpenter (Dr. W. B.), Announces Eozoön from Australia and Bavaria, and controverts the position of Profs. King and Rowney. Noticed in Amer. Jour. Sci. [2], xli., 406.

1866, Aug. King and Rowney (Profs.), "On the so-called Eozoönal rock," denying its organic character. Quar. Jour. Geol. Soc., Lond., xxii., pt. ii., 23. Their position is controverted in the same No. Their "Summary" is reproduced Amer. Jour. Sci. [2], xliv., 375.

1866. Gümbel (Dr.), "Occurrence of *Eozoön* in East Bavarian primitive rocks." Sitzüngsberichte d. K. Acad. d. W. in München, i., 1. Reproduced Quar. Jour. Geol. Soc., Lond., xxii., pt. i., p. 185. Noticed Amer. Jour. Sci. [2], xliii., 398. Confirmed by Carpenter, Proc. Roy. Soc., No. xciii., p. 508.

1867, May. Dawson and Logan. Describe new specimens of *Eozoön* from Tudor, C. W. Quar. Jour. Geol. Soc.

1867. Carpenter (Dr. W. B.), Reasserts organic nature of *Eozoön* in opposition to King and Rowney. Proc. Royal Soc., No. xciii., p. 503.

1867. Pusirevski (Prof.), Reports *Eozoön Canadense* at Hopinwara, Finland. Bull. Acad. St. Petersburg, x., 151. Noticed Amer. Jour. Sci. [2], xliv., 284.

1867, Nov. Dawson (J. W.), Notes on *Eozoön* from Tudor, C. W., from Long Lake and Wentworth, and from Madoc, with remarks by Dr. W. B. Carpenter. Amer. Jour. Sci. [2], xliv., 367. Republished Amer. Jour. Sci. [2], xlvi., 245 (Sept., 1868).

NOTE III., page 76.

As the reader may frequently desire to refresh his memory in reference to the order of superposition of the great groups of strata, the following table is appended for reference. The groups follow each other in the natural order of superposition.

CLASSIFICATION OF THE STRATIFIED ROCKS OF NORTH AMERICA.

Great Systems.	Systems.	Groups.	Localities.	Some leading Types of Fossils.
CENOZOIC.	Post Tertiary.	Terrace.	The existing surface.	Existing animals.
		Champlain.	Lake and river terraces.	Cave Bear, Hairy Elephant, etc.
		Glacial.	Deep-seated gravel and clays.	Primeval Man. Cave Bear and associates.
	Tertiary.	Sumter.	S. C.; Upper Missouri.	Extinct mammals.
		Yorktown.	Va.; "Bad Lands," Dakotah.	Great increase of mammals.
		Vicksburg.	Mississippi.	Orbitoides; Crocodiles.
		Jackson.	Southern Ala. and Miss.	Zeuglodon (whale-like).
		Claiborne.	Southern Ala.	Sharks; Sea-urchins; Shells.

CLASSIFICATION OF THE STRATIFIED ROCKS OF NORTH AMERICA—(*Continued*).

Great Systems.	Systems.	Groups.	Localities.	Some leading Types of Fossils.
MESOZOIC.	Cretaceous.	Later Creta. Earlier Creta.	Kansas; Nebraska. Montgomery, Selma, etc., Ala.	Ammonites; Sharks; Lizards. Angiospermous plants.
	Jurassic.	Wealden.	[Not known in America.]	Iguanodon and other Lizards.
		Oolite.	Jurassic strata, mostly west of Mississippi, extending into California.	Pterosaurs; Gar-pikes;
		Lias.		Ichthyosaurs; Plesiosaurs.
	Triassic.	Trias.	Kansas? Conn.?	Saurians and Labyrinthodonts.
PALEOZOIC.	Carboniferous.	Permian.	Kansas; N. C.	Coal plants and marine animals.
		Coal Measures.	Mid. and West. States.	Land plants; marine animals.
		Conglomerate.	Every where beneath Coal Measures.	Stems of drift-wood.
		False Coal Meas.	Ky.; Tenn.; W.Va.; O.	Land plants.
		Mountain Limestone.	West. States; St. Louis.	Crinoids; Corals; Chambered Shells.
		Marshall.	Cleveland; Burlington, Iowa; Michigan.	Chambered Shells; Brachiopods; Fishes.
	Devonian.	Chemung.	Southern New York.	Brachiopods.
		Hamilton.	Buffalo, N. Y.; Iowa City.	Brachiopods; Trilobites; Fishes.
		Corniferous.	Columbus, O.; Monroe, Mich.	Corals; Brachiopods; First Fishes.
		Oriskany.	Cumberland, Md.; Pa.; N. Y.	Brachiopods.
	Silurian.	Lower Helderberg.	Helderberg Mts., N.Y.; Me.; Mo.; O.; Mich.	Crinoids; Brachiopods.
		Salina.	Syracuse, N.Y.; Galt, C. W.; Mackinac, Mich.	Crustaceans.
		Niagara.	Niag. Falls; Chicago; Drummond's Isl.	Corals (Chain Coral, etc.).
		Cincinnati.	Cincinnati, O.; Nashville, Tenn.	Brachiopods; Corals (many small).
		Trenton.	N. Y.; Can.; Mich.; Tenn.	Orthoceratites; Gasteropods; Crinoids.
		Chazy.	E. side Kewenaw Pt.; Can.; N. Y.	Orthoceratites; Trilobites.
		Levis.	Quebec, Can.; Phillipsburg, N. Y.	Trilobites; Graptolites.
		Calciferous.	W. side Kewenaw Pt.; N. Y.; Canada.	Trilobites; Gasteropods; Corals.
		Potsdam.	Potsdam, N.Y.; St. Peter's R., Minn.	Trilobites; Brachiopods; Sponges.
		St. John's.	St. John's, N. F.; Georgia, Vt.	Trilobites; Orthis; Lingula; Cystids.
EOZOIC.	Huronian.	[Not yet subdivided.]	N. shore of L. Huron; Iron region of L. Superior; Some of the E. swells of the Appalachian chain.	Perhaps the Cambrian fossils of Europe are of this age.
	Laurentian.	Labrador. Lower Laurentian.	Adirondack Mts.; N. shore L. Superior; vast region north of St. Lawrence R., Canada.	Only fossil known is Eozoön. The American species is Eozoön Canadense. Plants probably existed.

NOTE IV., page 132.

The term Catskill group is employed in this connection in a sense greatly restricted from that in which the New York geologists originally employed it, since it has been shown that the principal portion of the so-called Catskill strata of the Catskill Mountains is really a prolongation of the Chemung rocks of the southern interior of the state; and for this reason, the original sense of the term is no longer admissible. The term thus stands as the designation of a series of strata which does not form a naturally restricted assemblage, and must drop out of use.

In parallelizing the "Catskill" (thus restricted) and the "Marshall" with the lower part of the Mountain Limestone of Europe, and at the same time suggesting their synchronism with the "Old Red Sandstone," I employ the latter term in its restricted and original sense, not as comprehending the whole recognized Devonian of the Old World. It is farther not unlikely that the parallelism ought to be restricted to the "Yellow," "White," and "Red" sandstones and conglomerates (Marwood and Petherwin beds) of the Old Red series, which, according to admissions made from time to time by Murchison and others, exhibit almost decisive affinities with the Carboniferous age. See Quar. Jour. Geol. Soc., London, vol. ix., p. 23.

NOTE V., page 139.

Allusion is here made to one of the mines at Lasalle, Illinois. This by no means exemplifies the greatest depth to which mining operations have been carried. The mine at Duckenfeld, in Cheshire, England, is probably the deepest coal mine in the world. A simple shaft was sunk 2004 feet to the bed of coal, and by means of an engine plane in the coal-bed, a farther depth of 500 feet has been attained, making 2504 feet to the bottom of the excavation. At Pendleton, near Manchester, coal is worked daily from a depth of 2135 feet; and the Cannel coal of Wigan is brought from 1773 feet below the surface. Many of the Durham collieries are equally deep, and far more extensive in their subterranean labyrinths.

The engine shaft of the Great Consolidated copper mines in Cornwall reaches the depth of 1650 feet, and the length of the various shafts, adits, and galleries exceeds 63 miles. Dalcoath tin mine, in Cornwall, is now working at more than 1800 feet from the surface. The famous silver mine of Valenciana, Mexico, is 1860 feet deep. The Hohenbirger mines in the Saxon Erzegebirge, near Freiburg, are 1827 feet deep, and the Thurmhofer 1944 feet deep. The depth of the celebrated mine of Joachimsthal, in Bohemia, is 2120 feet. The Tresavean copper mine in Cornwall is 2180 feet. The workings of the Samson mine at Andreasberg, in the Harz, have been prosecuted to the depth of 2197 feet. At Rörerbühel, in Bohemia, there were in the 16th century excavations to the depth of 3107 feet, *made before the invention of gunpowder*. The greatest depth to which human labor and ingenuity have as yet been able to penetrate is 3778 feet, in the old Kuttenberger mine in Bohemia, now abandoned. This depth, as remarked by Humboldt, is about eight times the height of the pyramid of Cheops or the Cathedral of Strasburg.

The deepest excavation in the United States is probably that of the Minnesota mine near Ontonagon, Lake Superior, which descends upon a copper-bearing lode to the depth of over 1300 feet. The Quincy mine at Hancock is 900 feet deep. The deepest mine in California is said to be the Hayward Quartz mine in Amador county, 1200 feet deep. The deepest excavations on the Comstock lode, Nevada, are 700 feet.

Mining excavations frequently extend from half a mile to a mile under the sea. In these gloomy subterranean and submarine passages, where, in some cases, one or two hundred feet of sea-water rest upon a slaty roof but three or four feet thick, the low moan of the waves can be continually heard above the miner's head, and in time of storms the howl becomes terrific and intolerable. The great adit for the discharge of the waters of the Gwennap tin mines in Cornwall exceeds 30 miles. In 1864, a tunnel 14 miles in length was completed in the region of the Harz mines, Brunswick, for the drainage of the district. A similar tunnel, 15 miles in length, designed for the drainage of the Freiberg district, has been in progress for several years. The Sutro tunnel, designed for the drainage of the mines located upon the Comstock lode, Nevada, is to be 19,000 feet in length, 12 feet wide, and 10 feet high, and will cost between four and five millions of dollars.

The great "tunnel" at Chicago, through which the city is supplied with pure water from Lake Michigan, is 10,567 feet long, five feet wide, and five feet two inches high, and at the shore extremity communicates with a vertical shaft 82 feet below the lake-level, and at the other extremity with a crib and shaft 66 feet below the lake-level.

NOTE VI., page 185.

As the *Archæopteryx*, or bird-reptile, is one of the most remarkable relics of the ancient world, and has but recently been brought to light, I append some references to sources of information upon the subject: Prof. Wagner first announced the discovery to the Royal Academy of Sciences of Munich in 1861; H. Herrman von Meyer described it in "Jahrbuch für Mineralogie," 1861, p. 561; Wm. H. Woodward, in "Intellectual Observer," Dec. 1862 (with plate); Prof. J. D. Dana, in "Amer. Jour. of Science and Arts," 2d ser., xxxv., May, 1863, p. 129, and "Manual of Geology," Appendix to later editions; Prof. R. Owen, in "Philosophical Transactions," cliii., part i., 1863, p. 33, pl. 1 to 4.

NOTE VII., p. 228.

I have given in the text the usual explanation of the phenomena of the glacial epoch. The theory of northern elevation, however, as the sole or principal cause of continental glaciers, has never been regarded, by many geologists, as completely satisfactory. Within a few years renewed attempts have been made to connect these phenomena with astronomical changes of a secular character.

At sundry epochs in the history of the world, agencies seem to have arisen which brought into existence and transported over considerable distances vast quantities of rounded pebbles and finer detrital materials. In some cases—as in the Niagara, Permian, and Upper Miocene periods—smoothed and striated rock-surfaces have been discovered, similar to those which are generally attributed to glacier action. In the intervening periods evidences of tropical temperature present themselves. The suggestion has therefore been made that more than once in the history of the world—perhaps at somewhat regular intervals widely removed—the northern portions of the continents have been visited by a reign of frost.

To account for these apparently secular phenomena, new investigations have been made upon the effects of the secular variations in the longitude of the equinox, the eccentricity of the earth's orbit, and the obliquity of the ecliptic. This is not the place to enter into an exposition of the discussions which have arisen. I may, however, simply explain the nature of the relation which subsists between terrestrial climates and the cosmical changes alluded to.

1. As to the variation in the eccentricity of the earth's orbit, it is evident that when the northern hemisphere has its winter in perihelion during the time of greatest eccentricity, the amount of glaciation must be considerably less than when the same hemisphere has its winter in aphelion during the time of greatest eccentricity.

2. As to the variation in the obliquity of the earth's axis, it appears that when the obliquity is greatest, the winter temperature of the polar regions can not be much severer than when the obliquity is least—since when the sun is below the horizon it is immaterial whether it be two degrees or ten below—while the summer temperature of the polar regions will be increased by the whole increase in the verticality of the sun's rays. The effect, therefore, of an increase in the obliquity of the earth's axis will be to diminish the average glaciation of the polar regions.

3. Suppose now the minimum glaciation of the polar regions, so far as due to obliquity, to occur at the time when the northern hemisphere experiences minimum rigors of climate through the effect of increased eccentricity; the conjunction of these two minima of cold in the north polar regions would, it is thought, remove the ice cap, and effect conditions of climate such as prevailed when Greenland, in the Miocene period, supported trees of tropical nature and luxuriance.

4. These conditions all reversed would produce a maximum of glaciation in the north temperate and polar regions such as evidently existed at the beginning of the Post Tertiary period.

It is probably within the power of physical astronomy to calculate the epochs at which these maxima and minima have occurred. It is, however, a problem of considerable difficulty, involving, as it does, the rate of precession of the equinoxes, the proper motion of the apsides, and the secular change in the obliquity of the ecliptic, none of which data are perfectly constant. According to recent determinations, the equinox completes a revolution in 25,868 years. The apsides move forward to meet the equinox, so that perihelion has the same longitude once in 21,066 years. The obliquity of the ecliptic returns to the same value in about 100,000 years.

M. Adhémar has based an explanation of the occurrence of glacial periods upon the climatic effects of the precession of the equinoxes alone. As the earth's axis is inclined to the ecliptic, the hemisphere which has its winter in aphelion is not only farther from the sun than the other hemisphere during its winter, but also experiences a winter having about eight days longer duration. The excess in the duration of its winter is partly caused by the slower motion of the earth on that side of the equinoxes which embraces the upper apsis, and partly also by the greater length of the path on that side. This hemisphere is therefore subjected to an excess of cold. For reverse reasons, the other hemisphere enjoys more than the mean warmth.

In consequence, however, of the gyration of the axis in a period which Adhémar takes at 21,000 years, it follows that at the end of 10,500 years the hemisphere which had been turned away from the sun at aphelion becomes turned toward him. In other words, the climatic inequalities of the two hemispheres become reversed. That hemisphere which for 10,500 years had been subjected to excessive glaciation, now enjoys excessive warmth, and that which had enjoyed excessive warmth is visited by excessive cold. There are, therefore, two great seasons for each hemisphere during the progress of the Annus Magnus, or Great Year. The summer has a duration of 10,500 years, and the winter an equal duration.

One thing farther should be remarked in connection with the accumulation of masses of ice and snow about either pole, from whatever cause the accumulation proceeds. Such an accumulation must necessarily change the position of the centre of gravity of the earth-mass. That centre must move toward the pole thus burdened. The fluent waters upon the earth's surface, free to adjust themselves in equilibrium about the centre of gravity, must change their distribution as the place of the centre of gravity changes. During the glaciation of the southern hemisphere, the waters will accumulate about the south pole; and during the glaciation of the northern hemisphere, they must accumulate about the north pole. These alternating accumulations of the waters are adequate, it has been calculated, to account for the last submergence of the northern hemisphere during the Champlain Epoch. The submergence of the southern polar lands is now in progress.

Should the connection of "the glacial period"—and of other glacial periods more ancient—with cosmical conditions, be satisfactorily established, it seems to me that we are here furnished with a hopeful means of giving greater precision to the calculus of geological time.

For detailed information on these questions, see Croll, Philosophical Magazine for August, 1864, and February, 1867; Transactions Geol. Soc., Glasgow, April, 1867; Revue des deux Mondes, 1847, etc.

NOTE VIII., page 287.

As the records of the flowing wells of oil springs constitute one of the most remarkable chapters of the history of Petroleum, I append here a list of them, made from personal examination and research. In this list, "Sub." stands for subdivision, "R." for range, "L." for lot, and "Con." for concession.

Former Flowing Wells at Oil Springs, township of Enniskillen, Ontario.

Depth in feet.		Yield in bbls.
3	Finn & Brown—S. E. part L. 17, Con. 1	Trifling.
104	Solis—Sub. 16, R. A., L. 18, Con. 2	600
108	Purdy—W. ½ L. 19, Con. 2	1000
115	Evoy Brothers—W. ½ L. 19, Con. 2	600
116	Jewry & Evoy—W. ½ L. 19, Con. 2	300
116	Fairbanks—Sub. 31, R. 5, L. 17, Con. 2	500
130	Campbell—W. ½ L. 19, Con. 2	200
132	Bennett Brothers	500
136	Chandler—Sub. 33, R. 2, L. 18, Con. 2	100
155	Jewry & Evoy—Same as above, bored deeper	2000
157	Sifton, Gordon, & Bennett—Sub. 2, L. 18, Con. 2	150
158	J. W. Sifton—Sub. 1, E part L. 18, Con. 2	800
158	Shaw—Sub. 10, R. B., L. 18, Con. 2	3000
160	Wanless—Sub. 6, R. E., L. 18, Con. 2	200
160	McLane—Sub. 2, E. part L. 18, Con. 2	3000
160	Ball—Sub. 3, E. part L. 18, Con. 2	250
160	Rumsey—Sub. 6, E. part L. 18, Con. 2	250
160	Whipple—Sub. 8, R. A., L. 18, Con. 2	400
163	Sanborn & Shannon—Sub. 13, R. C., L. 18, Con. 2	2000
163	Campbell & Forsyth—Sub. 12, R. C., L. 18, Con. 2	1000
163	Wilkes—Sub. 9, R. A., L. 18, Con. 2	2000
164	Bradley—Sub. 13, R. I., L. 18, Con. 2	3000
167	Webster & Shepley—E. part L. 18, Con. 2	6000
170	Leavenworth—Sub. 7, R. C., L. 18, Con. 2	500

Depth in feet.		Yield in bbls.
170	Culver—Sub. 7, R. C., L. 18, Con. 2	200
173	Allen—Sub. 32, R. 5, L. 17, Con. 2	2000
175	Barnes—Sub. 36, R. 5, L. 17, Con. 2	300
178	Petit—W. ½ L. 19, Con. 2	3000
180	George Gray—Sub. 32, R. I., L. 17, Con. 2	150
180	Holmes—Sub. 9, E. ½ L. 19, Con. 2	500
187	McColl—Sub. 37, R. 5, L. 17, Con. 2	1200
188	Swan—E. part L. 18, Con. 2	6000
196	Nelson—Sub. 29, R. 2, L. 17, Con. 2.	
212	Fiero—Sub. 1, R. 4, L. 19, Con. 1	6000
237	Black & Mathewson—Sub. 12, L. 17, Con. 1	7500

NOTE IX., page 396.

The thoughts embodied in this and the five following chapters were first shadowed forth by the author in the Michigan Journal of Education in 1860. They were more fully elaborated in the Ladies' Repository for November and December, 1863, and January, 1864. Many thoughts and conceptions which were then original appear to be now but the echoes of Mayer, Helmholtz, and others. This is particularly the case in reference to the doctrine of solar refrigeration. That doctrine, then entirely new to the writer, was put forth with much apprehension. The publication of Mayer's papers in Silliman's Journal (vol. xxxvi., p. 261; xxxvii., p. 187; xxxviii., p. 239, 397) in 1863 and 1864 afforded the writer the first exact basis for conclusions which he had already reached. The later researches of others have served to give a scientific sanction to statements which at one time might have been regarded as little more than vagaries of the imagination.

NOTE X., page 404.

Such, at least, is the generalization put forth by Müncke, Mrs. Somerville, and other physicists. It is apparently founded on reports of observations made by Scoresby and Parry in the Arctic Ocean, and by James Ross in the Antarctic. M. Charles Martins, however, a highly competent authority, denies that any such increase of temperature in the deeper arctic waters exists. Nothing of the kind was observed by le Contre-Amiral Coupvent des Bois in the voyage of the corvettes *Astrolabe* and *Zélée;* nor in the soundings made on the two voyages to Spitzbergen with the corvette *Recherche.*

On this subject, see Gehler's *Physikalisches Worterbuch*, t. vi., p. 1685; Somerville's *Connection of the Physical Sciences*, Am. ed., p. 245; Mayer, *Celestial Dynamics*, in Correlation and Conservation of Forces, p. 311; *Voyage en Scandinavie et au Spitzberg de la corvette la Recherche*, Géographie Physique, t. ii., p. 279; *Annales de Chimie et de Physique*, 3e Série, t. xxiv., p. 220, 1848; *Comptes Rendus*, t. lxi., p. 836.

INDEX.

THE END.

LOSSING'S FIELD-BOOK OF THE WAR OF 1812. Pictorial Field-Book of the War of 1812; or, Illustrations, by Pen and Pencil, of the History, Biography, Scenery, Relics, and Traditions of the Last War for American Independence. By BENSON J. LOSSING. With several hundred Engravings on Wood, by Lossing and Barritt, chiefly from Original Sketches by the Author. 1088 pages, 8vo, Cloth, $7 00.

LOSSING'S FIELD-BOOK OF THE REVOLUTION. Pictorial Field-Book of the Revolution; or, Illustrations, by Pen and Pencil, of the History, Biography, Scenery, Relics, and Traditions of the War for Independence. By BENSON J. LOSSING. 2 vols., 8vo, Cloth, $14 00; Sheep, $15 00; Half Calf, $18 00; Full Turkey Morocco, $22 00.

SMILES'S SELF-HELP. Self-Help; with Illustrations of Character and Conduct. By SAMUEL SMILES. 12mo, Cloth, $1 25.

SMILES'S HISTORY OF THE HUGUENOTS. The Huguenots: their Settlements, Churches, and Industries in England and Ireland. By SAMUEL SMILES, Author of "Self-Help," &c. With an Appendix relating to the Huguenots in America. Crown 8vo, Cloth, Beveled, $1 75.

WHITE'S MASSACRE OF ST. BARTHOLOMEW. The Massacre of St. Bartholomew: Preceded by a History of the Religious Wars in the Reign of Charles IX. By HENRY WHITE, M.A. With Illustrations. 8vo, Cloth, $1 75.

ABBOTT'S HISTORY OF THE FRENCH REVOLUTION. The French Revolution of 1789, as viewed in the Light of Republican Institutions. By JOHN S. C. ABBOTT. With 100 Engravings. 8vo, Cloth, $5 00.

ABBOTT'S NAPOLEON BONAPARTE. The History of Napoleon Bonaparte. By JOHN S. C. ABBOTT. With Maps, Woodcuts, and Portraits on Steel. 2 vols., 8vo, Cloth, $10 00.

ABBOTT'S NAPOLEON AT ST. HELENA; or, Interesting Anecdotes and Remarkable Conversations of the Emperor during the Five and a Half Years of his Captivity. Collected from the Memorials of Las Casas, O'Meara, Montholon, Antommarchi, and others. By JOHN S. C. ABBOTT. With Illustrations. 8vo, Cloth, $5 00.

ADDISON'S COMPLETE WORKS. The Works of Joseph Addison, embracing the whole of the "Spectator." Complete in 3 vols., 8vo, Cloth, $6 00.

ALCOCK'S JAPAN. The Capital of the Tycoon: a Narrative of a Three Years' Residence in Japan. By Sir RUTHERFORD ALCOCK, K.C.B., Her Majesty's Envoy Extraordinary and Minister Plenipotentiary in Japan. With Maps and Engravings. 2 vols., 12mo, Cloth, $3 50.

ALFORD'S GREEK TESTAMENT. The Greek Testament: with a critically-revised Text; a Digest of Various Readings; Marginal References to Verbal and Idiomatic Usage; Prolegomena; and a Critical and Exegetical Commentary. For the Use of Theological Students and Ministers. By HENRY ALFORD, D.D., Dean of Canterbury. Vol. I., containing the Four Gospels. 944 pages, 8vo, Cloth, $6 00; Sheep, $6 50.

ALISON'S HISTORY OF EUROPE. FIRST SERIES: From the Commencement of the French Revolution, in 1789, to the Restoration of the Bourbons, in 1815. [In addition to the Notes on Chapter LXXVI., which correct the errors of the original work concerning the United States, a copious Analytical Index has been appended to this American edition.] SECOND SERIES: From the Fall of Napoleon, in 1815, to the Accession of Louis Napoleon, in 1852. 8 vols., 8vo, Cloth, $16 00.

BANCROFT'S MISCELLANIES. Literary and Historical Miscellanies. By GEORGE BANCROFT. 8vo, Cloth, $3 00.

DRAPER'S CIVIL WAR. History of the American Civil War. By JOHN W. DRAPER, M.D., LL.D., Professor of Chemistry and Physiology in the University of New York. In Three Vols. *Vol. II. just published.* 8vo, Cloth, $3 50 per vol.

DRAPER'S INTELLECTUAL DEVELOPMENT OF EUROPE. A History of the Intellectual Development of Europe. By JOHN W. DRAPER, M.D., LL.D., Professor of Chemistry and Physiology in the University of New York. 8vo, Cloth, $5 00.

DRAPER'S AMERICAN CIVIL POLICY. Thoughts on the Future Civil Policy of America. By JOHN W. DRAPER, M.D., LL.D., Professor of Chemistry and Physiology in the University of New York, Author of a "Treatise on Human Physiology," "A History of the Intellectual Development of Europe," &c. Crown 8vo, Cloth, $2 50.

BARTH'S NORTH AND CENTRAL AFRICA. Travels and Discoveries in North and Central Africa: being a Journal of an Expedition undertaken under the Auspices of H.B.M.'s Government, in the Years 1849–1855. By HENRY BARTH, Ph.D., D.C.L. Illustrated. Complete in Three Vols., 8vo, Cloth, $12 00.

BELLOWS'S OLD WORLD. The Old World in its New Face: Impressions of Europe in 1867–1868. By HENRY W. BELLOWS. 2 vols., 12mo, Cloth, $3 50.

BOSWELL'S JOHNSON. The Life of Samuel Johnson, LL.D. Including a Journey to the Hebrides. By JAMES BOSWELL, Esq. A New Edition, with numerous Additions and Notes. By JOHN WILSON CROKER, LL.D., F.R.S. Portrait of Boswell. 2 vols., 8vo, Cloth, $4 00.

BRODHEAD'S HISTORY OF NEW YORK. History of the State of New York. By JOHN ROMEYN BRODHEAD. First Period, 1609–1664. 8vo, Cloth, $3 00.

BULWER'S PROSE WORKS. Miscellaneous Prose Works of Edward Bulwer, Lord Lytton. In Two Vols. 12mo, Cloth, $3 50.

BURNS'S LIFE AND WORKS. The Life and Works of Robert Burns. Edited by ROBERT CHAMBERS. 4 vols., 12mo, Cloth, $6 00.

CARLYLE'S FREDERICK THE GREAT. History of Friedrich II., called Frederick the Great. By THOMAS CARLYLE. Portraits, Maps, Plans, &c. 6 vols., 12mo, Cloth, $12 00.

CARLYLE'S FRENCH REVOLUTION. History of the French Revolution. Newly Revised by the Author, with Index, &c. 2 vols., 12mo, Cloth, $3 50.

CARLYLE'S OLIVER CROMWELL. Letters and Speeches of Oliver Cromwell. With Elucidations and Connecting Narrative. 2 vols., 12mo, Cloth, $3 50.

CHALMERS'S POSTHUMOUS WORKS. The Posthumous Works of Dr. Chalmers. Edited by his Son-in-Law, Rev. WILLIAM HANNA, LL.D. Complete in Nine Vols., 12mo, Cloth, $13 50.

CLAYTON'S QUEENS OF SONG. Queens of Song: being Memoirs of some of the most celebrated Female Vocalists who have performed on the Lyric Stage from the Earliest Days of Opera to the Present Time. To which is added a Chronological List of all the Operas that have been performed in Europe. By ELLEN CREATHORNE CLAYTON. With Portraits. 8vo, Cloth, $3 00.

COLERIDGE'S COMPLETE WORKS. The Complete Works of Samuel Taylor Coleridge. With an Introductory Essay upon his Philosophical and Theological Opinions. Edited by Professor SHEDD. Complete in Seven Vols. With a fine Portrait. Small 8vo, Cloth, $10 50.

DU CHAILLU'S AFRICA. Explorations and Adventures in Equatorial Africa: with Accounts of the Manners and Customs of the People, and of the Chase of the Gorilla, the Crocodile, Leopard, Elephant, Hippopotamus, and other Animals. By PAUL B. DU CHAILLU, Corresponding Member of the American Ethnological Society; of the Geographical and Statistical Society of New York; and of the Boston Society of Natural History. With numerous Illustrations. 8vo, Cloth, $5 00.

DU CHAILLU'S ASHANGO LAND. A Journey to Ashango Land: and Further Penetration into Equatorial Africa. By PAUL B. DU CHAILLU, Author of "Discoveries in Equatorial Africa," &c. New Edition. Handsomely Illustrated. 8vo, Cloth, $5 00.

CURTIS'S HISTORY OF THE CONSTITUTION. History of the Origin, Formation, and Adoption of the Constitution of the United States. By GEORGE TICKNOR CURTIS. Complete in Two large and handsome Octavo Volumes. Cloth, $6 00.

DAVIS'S CARTHAGE. Carthage and her Remains: being an Account of the Excavations and Researches on the Site of the Phœnician Metropolis in Africa and other adjacent Places. Conducted under the Auspices of Her Majesty's Government. By Dr. DAVIS, F.R.G.S. Profusely Illustrated with Maps, Woodcuts, Chromo-Lithographs, &c. 8vo, Cloth, $4 00.

DOOLITTLE'S CHINA. Social Life of the Chinese: with some Account of their Religious, Governmental, Educational, and Business Customs and Opinions. With special but not exclusive Reference to Fuhchau. By Rev. JUSTUS DOOLITTLE, Fourteen Years Member of the Fuhchau Mission of the American Board. Illustrated with more than 150 characteristic Engravings on Wood. 2 vols., 12mo, Cloth, $5 00.

EDGEWORTH'S (MISS) NOVELS. With Engravings. 10 vols., 12mo, Cloth, $15 00.

GIBBON'S ROME. History of the Decline and Fall of the Roman Empire. By EDWARD GIBBON. With Notes by Rev. H. H. MILMAN and M. GUIZOT. A new cheap Edition. To which is added a complete Index of the whole Work, and a Portrait of the Author. 6 vols., 12mo (uniform with Hume), Cloth, $9 00.

GROTE'S HISTORY OF GREECE. 12 vols., 12mo, Cloth, $18 00.

HALE'S (MRS.) WOMAN'S RECORD. Woman's Record; or, Biographical Sketches of all Distinguished Women, from the Creation to the Present Time. Arranged in Four Eras, with Selections from Female Writers of each Era. By Mrs. SARAH JOSEPHA HALE. Illustrated with more than 200 Portraits. 8vo, Cloth, $5 00.

HALL'S ARCTIC RESEARCHES. Arctic Researches and Life among the Esquimaux: being the Narrative of an Expedition in Search of Sir John Franklin, in the Years 1860, 1861, and 1862. By CHARLES FRANCIS HALL. With Maps and 100 Illustrations. The Illustrations are from Original Drawings by Charles Parsons, Henry L. Stephens, Solomon Eytinge, W. S. L. Jewett, and Granville Perkins, after Sketches by Captain Hall. A New Edition. 8vo, Cloth, Beveled Edges, $5 00.

HALLAM'S CONSTITUTIONAL HISTORY OF ENGLAND, from the Accession of Henry VII. to the Death of George II. 8vo, Cloth, $2 00.

HALLAM'S LITERATURE. Introduction to the Literature of Europe during the Fifteenth, Sixteenth, and Seventeenth Centuries. By HENRY HALLAM. 2 vols., 8vo, Cloth, $4 00.

HALLAM'S MIDDLE AGES. State of Europe during the Middle Ages. By HENRY HALLAM. 8vo, Cloth, $2 00.

www.ingramcontent.com/pod-product-compliance
Lightning Source LLC
LaVergne TN
LVHW020938110826
845150LV00004B/894

* 9 7 8 1 4 2 5 5 5 1 6 7 4 *